THE
MORE
EFFECT

Published by Make Good Media.
Jamison, PA
www.jeremyvictor.com
www.makegoodmedia.com

ISBN 979-8-9934174-0-0 (paperback)

First Edition

This book is a work of nonfiction. The ideas, practices, and strategies described are based on the author's experiences and research. They are provided for informational purposes only and do not constitute professional advice. The author and publisher disclaim any liability for actions taken in reliance on this material.

Cover design by Make Good Media
Interior design by Make Good Media
Written by Jeremy Victor – All original writing, no raw AI.
Edited by Jeremy Victor with ChatGPT and Gemini.

Printed in the United States of America

9 798993 417400

THE
MORE
EFFECT

Thriving in the Age of AI

A FIELD GUIDE FOR THE MODERNIZATION OF
HOW WE WORK, LIVE, & LEAD

JEREMY VICTOR

The history of the world is of conversations and handshakes.

Human *potential*
must keep it that way.

CONTENTS

O | OPPORTUNITY

Pillar R | RELATIONSHIPS

Pillar E – EXECUTION

SECTION 3 - THE HUMAN CODE

BACK MATTER

BLANK – FOR NOTES

BLANK – FOR NOTES

THE PEAK OF THE TREE

Have you ever looked, really looked, at the peak of a tall tree?
Its reach into the sky, leading the growth.
Upward. Into the unknown.

Fearless.

That reach is only possible through decades of storms, winds, and sun.
It is strength born from roots, tested by struggle, stretched by light.

The tree doesn't stop.
It keeps reaching. Higher.

Always higher.

So too with humanity.
Our growth, our leadership, comes not from ease but from struggle.
Not from shortcuts, but from the deep roots of courage, hope, and
optimism.

And in a world soon filled with machines and algorithms,
it is this pursuit, rooted, resilient, and reaching upward
that will keep us uniquely,

Human.

WHY "MORE" MATTERS NOW

> Whatever you can do, or dream you can, begin it. Boldness has genius, power, and magic in it.
>
> -Goethe

That quote has been my North Star as a futurist and as a leader who has navigated every major tech inflection point — from the birth of the Internet to today's AI revolution. It reminds me that progress isn't predicted; it is *initiated*: one bold experiment, one courageous question, one intentional act at a time. In today's AI-first, always-on world, that spirit of decisive beginnings matters now more than ever.

Algorithms can out-calculate and out-optimize us, but they cannot want. They cannot long for connection, feel gratitude, or summon courage when outcomes are uncertain. Those uniquely human traits fuel human potential and they must be cultivated with urgency in the AI age.

Growth mindset, as Carol Dweck framed it, teaches that abilities can be developed through effort and learning, pushing us toward *better*. For decades, that's been enough. Though today the accelerating pace of life is calling us to grow beyond better and to look deeper, stretch, and ask ourselves: *How do I Become More*?

Becoming More means growing **character** alongside competence, treating judgment, courage, and empathy as hard skills. It's the shift from performing better to *becoming someone* others can trust, follow, and build with.

Read with a beginner's mind and a builder's heart. Question habits that no longer serve you. Test the exercises in the messy laboratory of real meetings, late-night diaper changes, and quarterly targets, wherever your life happens. Share what you learn; wisdom hoarded is potential stranded.

Most of all, remember: **More** is not a mandate for endless hustle. It is a call to intentional growth: rooted in kindness, guided by curiosity, and measured by the positive ripples you create. If each of us commits to one courageous, hope-filled, optimistic act today, we will create a future where humanity and technology amplify one another rather than compete.

Thank you for opening this book and for opening yourself to the possibility that the best version of you, and of us, still lies ahead.

This is The MORE Effect.

A FRAMEWORK FOR GROWTH - PROFESSIONAL AND PERSONAL

The MORE Effect is a modern-day leadership field guide designed to propel your growth toward **Becoming More** both professionally and personally.

This book uses a simple framework. It starts with your:

- **Knowledge**: **what you're learning:** the information you gather and apply
- **Skills**: **what you're developing:** the proficiencies you sharpen and strengthen
- **Abilities**: **what you're leveraging:** the natural talents you bring
- **Mapping**: **how you're showing up:** the unique blend of knowledge, skills, and abilities that makes up your profile

You'll use this model throughout the book.

Section One introduces the origin story of The MORE Effect, the foundations of human-centric leadership, and discusses the evolving nature of managing and leading in the AI-first, always-on world. The stage upon which we are all living and leading.

Section Two is the **MORE Leadership Framework**. One by one, the 26 traits are revealed. Each chapter spotlights a single trait through stories, context, and practical exercises to learn, understand, and strengthen it. Along the way, two recurring sections will guide you:

- **How to Become More:** Reflection prompts and practices to help you assess where you are and choose your next step.
- **One MORE Thing**: A concise takeaway to anchor the lesson in everyday action.

By the end of each chapter, you'll know where you stand, what you've gained, and the next step forward.

Section Three integrates the **MORE Leadership Framework** into a practical Human Code for living and leading in an AI-first, always-on world. Together, the three sections provide context, practice, and integration: the why, the how, and the what.

Understanding Mapping - Your "Frequency of MORE"

Mentors early in my career introduced me to the idea of *mapping*. It stuck with me because it captured something essential: the unique blend of knowledge, skills, and abilities that make up your personal and professional DNA. Just as no two people share a fingerprint, no two people share the same mapping. We share the same traits, but each of us embodies them to different degrees. That variation creates your mapping, your unique **Frequency of MORE**. Think of it like a radio signal: we're all tuned to the same spectrum, but the volume, clarity, and resonance differ for each person. That's what makes your leadership voice distinct.

Why Mapping is the Starting Point for Growth

Mapping matters because self-reflection fuels growth. Without knowing where you stand, it's hard to know where to go. This framework builds awareness of your skills, tendencies, and opportunities for growth. Knowing your mapping shows where to adapt and grow. And remember, you are one whole person. Your personal and professional self are intertwined, and growth in one area often fuels growth in the other. True development, reaching your full potential, **becoming more,** only happens when you're able to be your full self at work and at home. The key is **awareness**. Understanding your personal mapping allows you to reflect on your current strengths and identify areas for growth. Reaching your full potential begins with understanding your mapping, the **unique Frequency of MORE** that only you carry. In the chapters ahead, you'll see how that mapping comes alive in practice – and how to **Become More.**

Questions for Reflection:

- Which leadership traits feel strongest in your current mapping?
- Which ones are faint signals you'd like to strengthen?

In Section 2, we'll explore 26 traits across four pillars — **Mindfulness, Opportunity, Relationships, and Execution** — that together form the foundation of your map.

SECTION 1

THE ORIGINS OF MORE

1 | INTRODUCTION - THE LEADERSHIP FIELD GUIDE FOR THRIVING IN THE AI-FIRST, ALWAYS-ON WORLD

Artificial intelligence is reshaping how companies operate, compete, and grow. The pace of work is no longer set by office hours; it's set by algorithms. A Slack message from the CFO at 5:45 a.m. on a Sunday, a customer who expects an answer in 30 seconds, a project launch pulled two weeks early, a board demanding efficiency and growth. This is the new reality leaders face: constant, 24/7, incoming demands. In this reality, leaders face a choice: compete with the machine on efficiency, or double down on the one thing technology can't replicate: our humanity.

This book is that guide — a modern leadership field manual rooted in human traits, values-based action, and the discipline to scale culture and performance in a world of constant change and ever-rising consumer expectations. It equips leaders, teams, and professionals at every level with the principles, mindset, and tools to foster collective intelligence, drive high-performance collaboration, and build careers and companies that not only scale, but endure the unknowns ahead.

I've honed and developed this framework over the course of my career, living through every major technology inflection point. Each time, I've seen the greatest outcomes come not from replacing people with technology, but from enhancing people's uniquely human capacities through it. This book is the result of those lessons.

What does this book mean by *More*?

- **The MORE Effect** is the outcome. It's the positive ripple created when people and teams consistently live the traits in this book and, in doing so, help others **Become More**.
- **The MORE Leadership Framework** is the discipline: a 26 trait system across four pillars: **Mindfulness, Opportunity, Relationships, Execution** – practiced with daily intent to thrive, grow, and lead with impact.

So why does it matter? Because as technology takes on more tasks, uniquely human capacities, empathy, intuition, creativity, kindness, and generosity, become what set us apart. Developing them isn't a one-time act; it takes reflection, intention, and steady work. And here's the good news: none of this requires a title. What it does require is discipline and commitment — the ongoing work of turning values into behavior and behavior into culture. It's how we close the gap between what we say and what we do when the world is chaotic. It's how adversity becomes a turning point.

This book is built for that kind of steady commitment. Not hacks. Not heroics. It avoids rigid models that no longer fit and instead meets leaders where they are — helping them move forward to **Become More**.

In an economy where everything fights for our attention, leadership becomes the work of amplifying human traits at scale. That's the work of this book. In the pages ahead, you'll see how ordinary moments become culture-shaping signals and how these traits turn complexity into momentum. To make sense of what's coming, we need a shared language for the work ahead. That begins with the messy, human origins of the idea, and one essential shift: from Better to **More**.

A FRAMEWORK MATTERS ONLY IF IT'S BUILT WITH THE FUTURE IN MIND.

2 | A WORD ON MORE

Unlocking the Potential Within Ourselves and Those Around Us

From Better to More

When I first started exploring the concept of *Better* as a leader and coach, it felt like a natural mindset for guiding teams and individuals. The idea of *Better* was clear and actionable: How do we improve the way we work, lead, and contribute? Whether on the field or in the workplace, it made sense to focus on continuous improvement, on being just a little better in every decision, every interaction. *Better* was the driving force behind how I encouraged those around me to grow.

But over time, I began to realize that *Better* wasn't enough. It focused on incremental gains and refinements to what already existed but didn't challenge us to go beyond the present. That's when the concept of **More** emerged for me, both as a leader and a coach. **More** isn't just doing things better; it's unlocking deeper potential in ourselves, in those we guide, and the community at large that we belong to.

As a leader and coach, embracing **More** means creating environments where individuals can expand their capacity, not just improve on what they already know, but grow in ways they never imagined. It's breaking through limitations: whether that's overcoming fear, challenging traditional ways of thinking, or encouraging bold experimentation. The pursuit of **More** isn't incremental change; it's seeking transformative growth – both personally and professionally.

Growth Mindset: A Foundation for More

The idea of a growth mindset, introduced by psychologist Carol Dweck, helps frame the pursuit of **More**. Dweck's work, especially her studies of children and students, shows us that intelligence and talent aren't fixed traits. Instead, they can be developed through hard work, desire, and perseverance. It's an idea that has transformed the way we think about potential in education, leadership, and personal growth.

Yet, as much as I believe in the growth mindset, there's a nuance that's important to acknowledge. The growth mindset focuses on improving what we already have: our existing talents, intelligence, and skills. But **More** is not refining the talents or knowledge we possess today. It's gaining new perspectives, expanding our horizons, and ultimately, **Becoming More** than we thought we could be. This mindset invites us to step into the unknown and unlock potential we may not even realize is there — in ourselves and those we lead.

Where a growth mindset helps us get better at what we already do, the philosophy of **More** takes us further. It is growing in ways we haven't yet imagined. **More** asks us to consider not just who we are, but who we can become.

What Does *More* Mean?

At its core, **More** is a philosophy of potential and possibility. It continuously unlocks new layers of ourselves and encourages those around us to do the same. It's not doing **More** for the sake of it. Instead, it's **Becoming More**: more intentional, more present, more human.

In your daily life, think of **More** as an invitation to explore beyond what's immediately visible. It takes you deeper: whether that's in the way we connect with others or in how we challenge ourselves to grow. In the professional world, **More** means looking at how we can elevate those around us, helping others find their unique strengths and contributions. It's not transactional. It is transformative.

More in Action: The Practical Side

Think of a moment in your life when you felt challenged to go beyond your comfort zone, when you discovered something about yourself that you hadn't seen before. That's **More** in action. Pushing past the surface to tap into your deeper potential.

In leadership, **More** might mean asking yourself: How can I help my team unlock their potential? How can I create an environment where they feel empowered to bring their full selves to the table? In your personal life, **More** might mean cultivating deeper relationships, being more present, more intentional, and more empathetic.

More is about showing up fully and committing to growth not just for yourself, but for the benefit of everyone around you.

More is Uniquely Human

In an age where technology and artificial intelligence are reshaping our lives, the pursuit of **More** takes on even greater significance. While technology can help us achieve greater efficiency, it cannot replicate the uniquely human traits that allow us to **Become More**, traits like courage, hope, optimism, gratitude, and kindness.

AI can assist us, but it cannot replace the uniquely human qualities that enable us to reflect, persevere, and create deep connections with others. These are the traits that allow us to navigate the complexities of life and help others unlock their potential, too.

The Power of Reflection: Your Journey to *More*

The key to unlocking **More** is self-reflection. Socrates taught, "The unexamined life is not worth living." When we take the time to examine our lives, our strengths, our challenges, and our growth, we gain the clarity needed to push beyond our limits.

Take a moment to reflect:

- Where in your life are you seeking **More**?
- Where do you feel stuck or limited?
- Who in your life could benefit from your help in unlocking **More** for themselves?

Self-reflection is more than personal growth. It is understanding how you can serve others more fully: whether that's your team, your family, or your community. When we unlock **More** in ourselves, we create the space for others to do the same.

Why I Believe the Endurance of Human Existence Depends on *More*

A futurist thinks in terms of decades, even centuries from now. Consider the accelerating pace of change: Moore's Law, phones with terabyte storage, the rise of artificial intelligence, and the prospect of artificial general intelligence simulating aspects of human behavior and decision-making.

But with all this advancement comes a challenge: If we don't act now to preserve the uniquely human traits that define us: our intuition, our empathy, our kindness, we could risk losing something essential. It's not difficult to envision a future where these qualities are overshadowed, where technology progresses so rapidly that we forget what it means to be human. You don't need to imagine the distant future to see the risk.

Today, the average person spends more than five hours a day on their phone. Algorithms are designed, constantly refined, and personalized to keep us watching longer, scrolling further, and consuming more. They're optimized for engagement, not for human flourishing. Every endless feed chips away at our attention, our presence, our ability to connect in real time. We need tools, and a framework, to fight this off, to reclaim our uniquely human capacities before they're eroded. The danger isn't just that we lose these traits within ourselves, but that we lose them in our communities.

Human connection is built on more than efficiency; it's built on trust, empathy, and shared understanding. If we allow technology to replace these values, we could face a future where community itself is lost. That's why I believe the endurance of human existence depends on one key word: **More**.

A Call to Action

The philosophy of **More** is a call to action. A reminder that as we embrace technological progress, we must also cultivate and protect the uniquely human traits that cannot be replicated. These are the traits that bind us together and enable us to create meaning, purpose, and connection in an increasingly complex world.

Ideas alone don't carry weight; they need roots. To see where **More** truly comes from, let me take you back. The framework wasn't born in theory; it was forged in obstacles, in lived experiences that demanded courage, hope, and optimism.

BETTER IMPROVES. MORE TRANSFORMS.

3 | ORIGIN STORY: FROM OBSTACLES TO THE MORE EFFECT

Turning Obstacles Into a Code for Becoming More

The MORE Effect didn't begin as a leadership model. It began with intestinal surgery at the Children's Hospital of Philadelphia (CHOP) when I was eight weeks old. It wasn't life-threatening, but it was my first obstacle, and where my nickname, Chop, was born. That scar and name, Chop, became a reminder that even when something small interrupts your life, you can choose how it defines you. For me, it became the start of a lifelong lesson in **resilience**.

Later, in my mid-twenties, when I struggled to find my footing, my mom gave the nickname new meaning. She inscribed its new definition:
Courage, **H**ope, **O**ptimism, and the **P**ursuit of All Three, inside the cover of a book. Those words became guiding principles for my life and the foundation of The MORE Effect.

When my mom reframed Chop as Courage, Hope, Optimism, and the Pursuit of All Three, she gave me more than a nickname. She gave me a personal

code. Those words became anchors I could return to when life felt chaotic
— a kind of early mapping before I even knew the word.

But the story of obstacles doesn't end there. My parents divorced when I
was two, and my dad relocated to California when I was eight. By the time
I was ten, he had lost his job, child support stopped, and he left my mom
alone with three kids under 18. Being abandoned scars you. But thankfully,
my mom and her best friend, my Aunt Joanne, gave me love, stability, and a
fighting chance. They taught me there's so much more to life than money.

Abandonment leaves a scar. But it also sharpened my **empathy** and my
drive to build stability for others. I learned that leadership isn't just about
providing direction — it's about creating a sense of safety and belonging
that I longed for as a kid.

Those early challenges shaped me. They helped me realize overcoming
obstacles wasn't just something I had to do. It became a core part of who I
am. I discovered that I was ***born to overcome***. Through those formative
years, facing obstacle after obstacle, I found the strength to persevere. I
held onto the belief that every obstacle was a stepping stone. That mindset,
my power to create better outcomes by giving more and pressing on, has
driven me every day. It's shaped my leadership approach, my relentless
pursuit of achievement, and my desire to always **Become More**.

I invite you to reflect on the moments that made you. Like my childhood
experiences, what challenges have helped you grow and fueled your
purpose? For me, those hard moments became a foundation built on
courage, hope, optimism, and the pursuit of all three.

Adversity makes us who we are.
-Nick Sirianni

Each of these experiences carried a lesson — resilience from surgery, hope
and perseverance from my mom's words, empathy from being left behind.
At the time, I didn't see them as a framework. But looking back, they were
the first building blocks of what became this book.

From those early lessons came human-centric leadership. What started as survival became a framework, the **MORE Leadership Framework,** a guide for how anyone can lead, no matter their title.

These experiences taught me that the difference between coping and thriving was the choice to **Become More** – a lesson that shaped this entire book.

Questions for Reflection

- What early obstacles have shaped how you lead today?
- Which challenge taught you the most about who you are?
- How might those experiences serve as building blocks for your own framework of **Becoming More**?

OBSTACLES AREN'T DETOURS; THEY'RE THE PATH.

4 | THE MORE LEADERSHIP FRAMEWORK

Leadership today is not a title or a position. It's how you show up and how you contribute, in moments of pressure, in moments of decision, and in the quiet moments no one sees. The modern workplace requires everyone, at every level, to lead. And with another major technological inflection upon us, that truth has never been more urgent.

Introducing the **MORE Leadership Framework** now is important to provide context for the remaining chapters in Section One. The twenty-six traits being introduced and expanded upon in depth in Section Two are the disciplines of modern human-centric leadership. They are not abstract ideas. The traits are practices; daily choices meant to shape how you lead yourself and others, the uniquely human work of modern leadership.

I didn't choose these pillars and traits at random. They emerged over decades of my life inside and outside of work: leading, coaching, and volunteering. Over time, patterns repeated: I noticed the leaders and teams that leaned into these disciplines outperformed those that didn't. Each time I helped someone uncover their potential, the framework itself became more clear.

To make the traits memorable, the **MORE Leadership Framework** organizes the traits into four foundational pillars:

- **M | Mindfulness**: the ability to be present, steady, and aware.
- **O | Opportunity**: the conviction that the future is open and can be built.
- **R | Relationships**: the trust that binds people together and allows them to thrive.
- **E | Execution**: the discipline of moving with purpose and turning intent into impact.

These four pillars capture the balance leaders need today: attention and presence to see clearly (Mindfulness), optimism to pursue possibility (Opportunity), connection to bring others with you (Relationships), and discipline to make it real (Execution).

Within each pillar are the traits that keep leadership human-centric in an AI-first, always-on world. They are how you stand out, how you endure, and how you help others **Become More**.

Here is the full roadmap of traits you will learn and explore:

- **M | Mindfulness**: Mindfulness, Self-Awareness, Emotional Intelligence, Gratitude, Resilience, Intuition, Balance.
- **O | Opportunity**: Optimism, Hope, Perseverance, Generosity, Motivation, Curiosity, Vision.
- **R | Relationships**: Empathy, Social Awareness, Listening, Conflict Resolution, Sincerity, Kindness.
- **E | Execution**: Courage, Critical Thinking, Decision-Making, Creativity, Adaptability, Temperance.

MINDFULNESS	OPPORTUNITY	RELATIONSHIPS	EXECUTION
M	O	R	E
Mindfulness	Optimism	Empathy	Courage
Self-Awareness	Hope	Social Awareness	Critical Thinking
Emotional Intelligence	Perseverance	Listening	Decision-Making
Gratitude	Generosity	Conflict Resolution	Creativity
Resilience	Motivation	Sincerity	Adaptability
Intuition	Curiosity	Kindness	Temperance
Balance	Vision		

As you explore these traits of modern human-centric leadership, remember they are not fixed qualities but disciplines to be practiced and cultivated. Taken together, they form your field guide for thriving today and preparing for tomorrow.

Any map is only as useful as the people it serves. All leadership fails when it centers the individual over the whole. One of the foundational principles of leadership comes from Zig Ziglar, "You can have everything in life you want, if you will just help enough other people get what they want." This principle prioritizes others before self and anchors the **MORE Leadership Framework** to a simple premise: no one is bigger than the community.

ADVERSITY GAVE THE FRAMEWORK ITS ORIGIN; OPTIMISM GAVE IT WINGS.

5 | THE FIRST RULE OF MODERN LEADERSHIP

No One Is Bigger Than the Community

In today's world, where every moment can be posted online for likes, it's easy to mistake attention for achievement. Individualism is celebrated, and the chase for approval can quickly turn into serving one's ego. In that environment, it's no surprise we lose sight of the first rule of leadership: no one — no single person — is bigger than the community they are part of. By community, I don't just mean society at large. I mean the team, organization, or mission you belong to — the whole that outlasts any single individual. Mission comes first. Whether you're leading a company, managing a small team, or volunteering in your neighborhood, understanding this truth is key to **Becoming More**.

The Ego Trap

I've spent years observing how ego can sabotage even the most promising efforts. Ego tells us that success is individual. That to be more, we need to outshine others. But ego is a trap. It isolates us, creating a false narrative that we can only achieve greatness by standing alone at the top. WeWork's implosion under Adam Neumann, or Theranos collapsing under Elizabeth Holmes' deception, are reminders of how quickly ego can undermine progress and destroy trust. I've seen it up close too — like the time a colleague took credit for a project I initiated. Ego destroys teams, disrupts cultures, and undermines potential.

To lead, your role becomes to support the success of others. You'll never be bigger than the community that supports you. And if you try to be, you'll find yourself not just disconnected from others but disconnected from the primary purpose of a leader. Whether you're an emerging leader, an entrepreneur, or someone still figuring out your path, this is a lesson to learn early: the measure of success isn't solitary achievement, it is the collective achievement you inspire in those around you.

Leadership in the Age of AI

In the age of AI, it is important leadership remains rooted in people and community — creating spaces where empathy, collaboration, and trust are as important as contribution and productivity. Machines can scale output, but only leaders can scale belonging. It begins with accountability and building a culture where people feel seen, valued, and part of something larger than themselves.

When we lead for the whole, individual potential compounds. I've seen it again and again: aligned teams act with urgency, learn faster, and help each other without pause. The deeper the principle that one's success is born from others is instilled in your team, the more selfless the whole becomes. It's a self-fulfilling prophecy. Through shared accountability and outcomes, trust grows, generosity expands, and connections strengthen.

The future belongs to leaders who understand that community is the force multiplier - "help enough other people get what they want." Individual wins matter, but they matter most when they contribute to the collective good. Moving forward, effective leadership won't be about standing alone at the top but about building a foundation of shared success. The challenge for leaders is to foster cultures where empathy and collaboration are as valued as productivity, and where every individual feels seen as part of something larger. That's why later in this book we'll explore traits like empathy, listening, and kindness — the daily practices that make community thrive. By understanding that the strength and bond of the community is the true "force multiplier," we can ensure that future innovation and achievement are built on a foundation of purpose, trust, and shared accountability.

Community thrives only when the systems around it evolve. That's why an eye toward the future and what's possible becomes essential for modern leadership.

**YOU HAVE ONE OUTCOME AS A LEADER,
TO SUPPORT THE SUCCESS OF OTHERS.**

6 | THE FUTURE BELONGS TO THOSE WHO MODERNIZE

How to Replace Rigid Systems with Future-Ready Ones

We live in a world that, in many ways, hasn't caught up with the speed of our own progress. For example, election results in the U.S. are gathered through fifty different state systems, all feeding into the Associated Press, which then calls the outcome. That approach isn't just archaic. It is rigid and resistant to change, revealing that many structures we rely on every day are long overdue for modernization.

The same pattern shows up in business: Blockbuster clung to late fees while Netflix reinvented home entertainment; Kodak resisted digital photography and watched its market vanish; Sears doubled down on catalogs while Amazon built the future of retail. Each failed not because the future wasn't visible, but because rigidity blinded them to it. This rigidity is a hallmark of outdated systems. When organizations (or individuals) become fixed and remain unyielding to change, they fail to meet the needs of a constantly evolving world.

Modernization, a type of strategic thinking, dismantles this rigidity, creating systems that adapt, respond, and grow alongside us.

Crossing "The Chasm" of Modernization

The Technology Adoption Life Cycle, illustrated by Everett Rogers, highlights a critical phase known as "The Chasm," a gap many systems, products, and businesses struggle to cross. This divide represents the transition from outdated, rigid ways of operating to flexible, future-ready approaches. Modernization serves as the bridge across this chasm. It enables us to reshape entrenched systems and thinking, giving ideas the agility they need to adapt, evolve, and thrive. Think of modernization as transforming stagnant processes into adaptive, resilient ones.

Modernization: Assume It Exists, Then Create It

Modernization is a mindset shift that challenges rigidity. It begins by envisioning a future where obsolete processes are replaced with agile, adaptive systems. That work in ways you would expect them to in a modern world. Imagine: *assume that the future state already exists*, and then working backward to create it. This approach makes modernization a practice — a skill of a leader — proactive rather than reactive.

As reflected in the idea "Assume It Exists, Then Create It," *modernization requires forward-thinking.* Leaders who excel don't just adopt new technologies. They question existing systems and ponder, "What if?" This brings flexibility into our thinking and allows one to challenge existing constructs that have become too rigid to serve today's demands.

Modernization Principles

- **Curiosity:** Continuously ask *"What if?"*
- **Continuous Learning:** Never assume today's skills are enough.
- **Adaptability:** Treat change as normal, not exceptional.
- **Vision:** Envision a better system, then work backward to build it.
- **Courage:** Challenge entrenched practices even when it's uncomfortable.

Modernization in Daily Life and Work

Modernization isn't limited to abstract systems; it's something we experience in our daily lives. From digital calendars replacing physical planners to AI-driven decision-making, modernization ripples through our routines. But real transformation happens when we adopt a modernized mindset, questioning each process: *How can we make this better, more efficient, and future-ready?*

In the workplace, this mindset distinguishes high performers. Leaders who thrive today reject rigidity, seeing modernization as a continuous practice of optimizing every step. They recognize that innovation comes not from preserving the status quo but from constantly reimagining what's possible.

Tools and Mindset: Building the Future at the Speed and Pace of Change

Today's tools, AI, data analytics, and automation, support almost any ambition. Yet tools alone aren't enough; they require the right mindset. Modernization is about using available resources to dismantle rigidity and build systems that evolve.

Returning to the Technology Adoption Life Cycle, modernization is about moving forward with intentionality, ensuring innovations cross "The

Chasm" from early adoption to mainstream integration. The tools we choose and the mindset we adopt determine how effectively we build the future at the speed of change.

Developing a Modernization Mindset

Leaders of tomorrow won't be defined by titles or technical expertise alone. They'll be defined by how they think—by their ability to modernize. A modernization mindset sees change as essential for growth. It rejects rigidity and embraces adaptability as a core strength, both in life and at work. This mindset reshapes how leaders show up: asking better questions, challenging outdated systems, and creating the conditions for people to thrive in environments that never stop evolving. To modernize is to make flexibility a discipline, not a reaction.

Modernization is adaptability in practice, creativity in motion, and vision with discipline. These are traits we'll explore in detail later, but together they make modernization not just a tactic, but a way of leading. For those who embody it, modernization becomes more than a tactic—it's a principle for thriving in an AI-driven world. It's what enables leaders to cross the chasm from rigidity to resilience, and to innovate at the intersection of technology and humanity.

The Future Belongs to Those Who Modernize

As we navigate a world defined by rapid change, the ability to modernize isn't just a skill. It's a mindset. Embracing modernization means recognizing that today's systems might be outdated tomorrow. By consistently assessing and adapting, we break through rigidity and shape progress.

The future belongs to those who dare to imagine, innovate, and modernize, not just for today, but with a vision for tomorrow. **This is the mindset that separates leaders who thrive in the AI-first, always-on world from those who are left behind.** Yet modernization alone isn't enough. Without trust, nothing else we build will hold.

A FUTURIST CREATES WHAT'S YET TO BE SEEN.

7 | LEADING WITHOUT LOOKING

Trust and Modern Work Ethics in the AI-Driven Age

Trust is a delicate thing. It begins as an assumption, a leap of faith between employer and employee. When someone is hired, there's an unspoken agreement: *I trust you'll do the job with integrity and purpose; you trust I'll provide the environment and tools you need to thrive.* Trust isn't permanent. It isn't set in stone on Day One. It's more like a bond, one that's strengthened or weakened with every interaction, every deliverable, every choice.

In remote and hybrid work environments, this dynamic becomes even more critical. Without the proximity of shared office spaces, where you can feel the energy of others or lean over for a quick chat, trust becomes the oxygen of the relationship. It fuels everything.

So, how do we make sure the bond of trust isn't just assumed but actively nurtured, every single day?

Trust: The Assumption at Hiring

When a new hire joins a team, trust starts as a clean slate:

- Employers trust that each person will bring their full effort, creativity, and reliability to the table.
- Employees trust that the organization will be fair, transparent, and supportive in return.

This mutual trust is the handshake, virtual or otherwise, that gets things started. But assumptions alone won't hold. What happens next is where the real work begins.

If trust is the starting line, then ethics are the daily steps that keep you moving forward. Without them, trust collapses under the weight of ambiguity, inconsistency, or neglect. As companies issue return-to-office mandates and struggle with hybrid work models, we're witnessing a clash between old and new ways of thinking about work.

Today's employees see themselves on equal footing with their employers, armed with skills, options (the gig economy), and a clear sense of their value. This shift demands a new framework for trust. One that acknowledges this evolved relationship and provides guardrails for both sides. I call this framework **Modern Work Ethics**.

Modern Work Ethics: Forging Trust Daily

Rather than relying on oversight and control, Modern Work Ethics emphasizes mutual respect, clear communication, and high-character behavior from all parties. I learned that trust isn't built by policies alone, but by daily behaviors. Here are the core practices my team embraced to turn fragile trust into durable partnership:

- **Clarity & Communication**: Be clear, consistent, and human. Define expectations and goals up front; silence or ambiguity erode trust faster than anything else.

- **Accountability**: Own your role. Do what you said you'd do, when you said you'd do it. And when you stumble, admit it, fix it, move forward.
- **Boundaries & Balance**: Respect the line between personal and professional. Protect time off and acknowledge that life happens. Trust strengthens when people are seen as whole humans.
- **Recognition**: Celebrate wins, big and small. A simple, genuine "thank you" builds loyalty and momentum.
- **Flexibility with Responsibility**: Remote and hybrid work thrive on freedom, but freedom must be matched with discipline. Flexibility only works if accountability keeps pace.
- **Mutual Investment:** Trust is reciprocal. Employees and employers alike must invest time, energy, and care to make the relationship work.

Ground Rules in Practice

To make these principles real for my own team, I drafted a simple set of ground rules. We shared them, discussed them openly, and held each other accountable.

- **Presence & Availability:** Don't just show up: engage. Cameras on, calendars updated, Slack statuses clear. If you're in the room, be in the room. If you're out, share why. Remote or hybrid, show respect for the people you're working with. Silence is remembered and questioned.
- **Contribution & Productivity:** Do work that matters. Determine the highest and best use of your time. Don't waste energy on "busywork" that can be handled by AI. When you look at yourself in the mirror, if you're not proud of what you contributed, if it was not your best work, ask yourself why and rethink how you're using your time. Innovation beats inertia every time.
- **Work-Life Balance:** Respect time off. Yours and others. Boundaries may be blurred, but they're still worth protecting. Take PTO. Encourage your team to do the same. If you're giving maximum contribution to your role, it's important you spend time resetting and taking care of your mental and physical health.

These ground rules demonstrate that trust isn't a given; it's earned through consistent, high-character, ethical choices.

The First Test of Leadership

Trust isn't static. It's an active bond — strengthened or weakened — with every action. Leaders who understand this make establishing trust their first and most important responsibility. Without trust, high performance will never be achieved.

By embedding ethics and establishing ground rules into daily work, leaders transform fragile assumptions into resilient bonds creating a culture where both sides feel valued and respected. And in the age of AI, where algorithms can optimize output but never build relationships, that's the work that will set you apart. When trust and ethics are alive, teams have a foundation upon which to work smarter together. That's the promise of collective intelligence.

THE FIRST TEST OF LEADERSHIP IS TRUST.
TRUST IS EARNED WHERE RESPECT IS GIVEN.

8 | HARNESSING COLLECTIVE INTELLIGENCE: LEADERSHIP IN THE MODERN WORKPLACE

How Modern Leaders Can Unlock Team Potential, Foster Innovation, and Make Better Decisions

Leadership has always been about decisions. What's changed is the system around them. Complexity has multiplied, AI now informs every choice, and employee expectations have shifted. In this environment, the old command-and-control model falters. Leaders who thrive are those who turn trust into collaboration, and collaboration into higher performance, better outcomes and modern workplaces.

Organizations now require leaders to harness the intelligence of many, not just the expertise of one. This shift isn't theoretical; it's essential. Companies that actively promote collaborative decision-making consistently outperform their peers. Leaders who embrace **collective intelligence,** the shared wisdom and problem-solving capacity of teams, can drive better decisions, unlock innovation, and create workplaces where teams of diverse individuals thrive.

Why Leaders Must Embrace Collective Intelligence

Traditional command and control leadership models were built for a world that no longer exists. The modern business landscape is defined by:

- **Increased Complexity** – No single leader can have all the answers. The problems we face today require diverse perspectives and cross-functional expertise to solve effectively.
- **The Rise of AI and Data-Driven Decision-Making** – Leaders must integrate AI and real-time insights without losing the human element of judgment, ethics, and emotional intelligence.
- **Changing Workforce Expectations** – Employees today want to be part of the decision-making process. They expect leaders to foster collaboration, transparency, and inclusivity rather than command-and-control leadership.

What Is Collective Intelligence?

At its core, collective intelligence is the ability of a group to make better decisions than any individual could alone. When teams operate with openness, trust, and structured collaboration, they consistently generate more innovative ideas, stronger solutions, and better business outcomes.

Why Collective Intelligence Works

- **Diversity of Thought** – Teams with varied backgrounds, skills, and perspectives produce more creative solutions than homogeneous groups.
- **Cognitive Load Distribution** – No single person can process everything. Sharing knowledge reduces blind spots and helps make better-informed decisions.
- **Shared Accountability** – When people are part of the decision-making process, they take ownership of the outcomes. This increases motivation and engagement.

How Leaders Can Foster Collective Intelligence

1. **Create Psychological Safety** - For teams to engage in honest, productive collaboration, they need to feel safe voicing their opinions. Without psychological safety, collective intelligence dies.

Leaders should:
- **Encourage curiosity and dissent** – Invite differing viewpoints and treat disagreement as a sign of engagement, not defiance.
- **Reward learning over perfection** - Frame mistakes as opportunities for learning.
- **Listen more than you speak** – Demonstrate that every idea is valued and appreciated.

Tip: Ask yourself: "Am I creating an environment where my team feels safe challenging my ideas?"

2. **Use Open Dialogue & Debate to Surface the Best Ideas** - Effective collective intelligence doesn't mean everyone blindly agrees. It means people debate constructively to uncover the best ideas.

Leaders should:
- **Normalize disagreement as healthy** – Frame debates as a way to strengthen ideas, not as personal conflicts.
- **Facilitate structured discussions** – Use formats like "Red Teaming" (where a small group challenges a dominant perspective) to pressure-test decisions.
- **Ask expansive, open-ended questions** – Instead of asking, *"Do you agree?"* ask, *"How could this idea fail?"* or *"What are we missing?"*

Tip: Great leaders don't impose their views; they create space for smarter solutions to emerge.

3. **Leverage AI to Enhance, Not Replace, Human Decision-Making** - Collective intelligence is amplified when humans and AI work together. Leaders should use technology to surface insights but apply human judgment to make the final call.

Leaders should:

- **Use AI for pattern recognition** – AI can analyze vast amounts of data and identify trends faster than humans.
- **Let humans provide context and ethics** – AI lacks empathy, intuition, and ethical reasoning. Leaders must apply these to decision-making.
- **Make collaboration frictionless** – Use digital collaboration tools (like real-time polling, AI-driven knowledge hubs, or digital whiteboards) to ensure ideas are captured and shared efficiently.

Tip: Use AI as a teammate, not a replacement for human judgment.

4. **Shift from Leader as "Dictator" to Leader as "Facilitator"** - Command-and-control leadership models placed leaders at the top as decision-makers giving orders. A top-down approach limits performance and often stifles collaboration. The Marines' principle of Commander's Intent proved that when everyone understands the mission, decisions can be distributed without losing accountability.

Modern leadership requires:
- Asking clarifying questions to ensure understanding of desired outcomes, not just giving orders.
- Making knowledge transparent instead of hoarding information.
- Facilitating inclusive decision-making instead of deciding in isolation.

Practical ways to make this shift:
- Rotate decision-making roles. Use roundtables where different team members lead discussions.
- Crowdsource before committing. Gather diverse input before finalizing high-stakes decisions.
- Build, don't block. Adopt a "Yes, and..." mindset to expand ideas instead of shutting them down.

A leader's job is no longer to be the smartest person in the room. It is to make the room smarter.

The Future of Leadership Is Collective

Collective intelligence is not just about making better decisions. It is about building stronger teams. When leaders harness the collective wisdom of their people, they create:

- More innovative solutions to complex problems
- A more engaged workforce that feels heard and valued
- A culture of continuous learning and adaptability

The leaders who succeed in the future won't be the ones with all the answers. They'll be the ones who know how to unlock the intelligence of those around them.

So, the real question isn't *if* you should embrace collective intelligence. It's *how fast* you can start. And once the best ideas surface, the next challenge is making systems flexible enough to fit the lives of modern knowledge workers.

WHY BE ALONE AT THE TOP? BE SMARTER, TOGETHER.

9 | THE NEW FRONTIER OF WORK-LIFE BALANCE

How Modern Parenting, Flexibility, and Empathy Will Shape the Future of Work

Introduction: The Blurring Lines Between Work and Life

For decades, we've chased the elusive "work-life balance," a concept that implies we could compartmentalize our personal and professional lives. But the world has changed. Now, after the pandemic, the boundaries between work and life are more blurred than ever, especially for modern parents. Flexible work arrangements, family responsibilities, and technology have made it increasingly difficult to separate the two.

As a leader who has worked with diverse teams across different industries, I've learned that today's employees are not just asking for flexibility. They're asking for understanding. Modern parenting and family dynamics are no longer separate from work; they are intertwined.

The future of work is more than flexibility. It begins with empathy and creating environments where work and life can coexist harmoniously.

Personal Story: The Lesson I Learned as a Leader

In my experience as a leader, one thing became clear over time: employees are no longer simply looking for a job that pays the bills. They are looking for work environments that understand their personal lives. At the same time, balance without achievement may leave employees feeling hollow. Many employees want to know they're on a winning team, that their work and effort are producing real results for the company and, ideally, the world. Pay matters. Time off matters. But pride in contribution matters just as much. A leader's role isn't to balance someone's life for them, but to create the space and conditions where they can balance it themselves while still feeling they're part of something that wins.

I've had countless conversations with employees who weren't just asking for remote work or flexible hours to avoid a commute. They were asking for it because they were balancing personal commitments, from childcare to eldercare to their own well-being, along with the responsibility they felt to their job. They sought a best-of-both worlds situation. What they wanted was an employer, really me as their supervisor, who would listen and understand that, deep down, they were simply asking for help.

Through these discussions, I realized there is no longer a clear separation between work, life, and family. These spheres overlap, and employees are seeking workplaces that support this reality. It wasn't always easy, especially in environments where strict policies frowned upon flexibility. But I knew that listening to employees was the key to creating a more supportive workplace. Over time, I discovered that when employees felt their personal needs were respected, they became more engaged and loyal to me, and by extension, the company.

It was through these conversations that I realized today, there is no true separation between work, life, and family. For me, that's the essential takeaway that must be woven into the future of work when we think about the policies that govern it.

Work and Life: The Evolution Across Generations

The shift in work-life balance isn't just about individual preferences; it's a generational trend. Baby Boomers and early Gen Xers grew up in an era where the 9-to-5 workday was the norm, and "work-life balance" meant going home after work to spend time with family. But as we moved into the digital age, those lines began to blur.

For Millennials (Gen Y) and especially Gen Z, work and life are no longer seen as separate entities. These generations prioritize flexibility, personal fulfillment, and meaningful work over rigid structures. They understand that life happens in between meetings, emails, and deadlines and they expect their employers to understand that too.

According to a McKinsey report[1], 87% of workers offered flexible work arrangements take advantage of them, signaling that flexibility is no longer a perk. It is a necessity for retaining talent. Companies that fail to adapt to this evolving landscape risk losing out on the next generation of top performers.

Flexibility, though, isn't one-size-fits-all. In industries like manufacturing or healthcare, daily in-person collaboration is essential, and often required for early-career or newly hired employees. Remote tools help in emergencies, though physical presence still drives performance in those settings. This reality pushes leaders to redefine what flexibility means in their context, in order to attract top performers while matching the demands of the work itself.

Companies Leading the Way

Some U.S. companies are already embracing this shift. Salesforce, for example, introduced its "Success From Anywhere" model, giving employees the freedom to work wherever they feel most productive. True flexibility goes beyond location; it gives people the autonomy to manage their time and responsibilities. Salesforce has shown that flexibility is not only achievable but also a driver of employee engagement and productivity.

Patagonia is another leader in this space. Known for its on-site childcare and family-friendly policies, Patagonia has created a culture that recognizes the importance of supporting employees as whole people, not just as workers. Their approach has fostered deep loyalty among employees, many of whom stay with the company for years because they feel understood and valued. In Patagonia's case, another equally important factor to employee loyalty is Patagonia's larger mission — a visible, decades-long commitment to the environment and sustainability offers purposeful, mission-driven work.

These companies understand that flexibility and empathy are key to building the future of work. They have moved beyond traditional approaches to work-life balance and are creating environments that encourage employees to bring their whole selves to work. This isn't just about remote work; it's about building workplaces that reflect the realities of modern life and families.

The Future of Flexibility: How Companies Could Adapt

Looking ahead, the companies that will thrive in the future will be those that recognize the need to deeply understand the ever-evolving needs of their workforce, both inside and outside of work. For knowledge workers, flexible hours and remote options will be table stakes in this conversation.

We are already seeing how Gen Y and Gen Z are prioritizing life over work. Ignoring this fact by holding onto outdated policies will soon no longer be an option.

At the same time, it should be stated that Gen Y and Gen Z will also need to understand that advancement often still requires sacrifice such as the need for extra hours, stepping up when the team needs it to meet the needs of the day. Leaders need to balance empathy with clarity, making sure employees know that growth comes from both balance and contribution.

At the core, employees don't just want balance, they want their work to matter, to know their contributions have impact beyond the task itself.

The need for allowing employees to balance their professional and personal lives will become the new normal, essential for retaining employees and thriving in the modern workplace.

By exploring new ways to support employees, beyond the traditions of the past, companies can lead the way in shaping a future that works for everyone. The shift towards flexibility is a response to trends, recognizing the changing dynamics of family life and the pressures employees face. Companies that adapt will be well positioned for the future, creating environments where employees can thrive both at work and at home.

A Vision for the Future of Work-Life Balance

The future of work will no longer be rigid policies and clocking in from 9 to 5. It will evolve and become an understanding that work, life, and family are no longer separate compartments of our lives. They are interconnected, and companies that embrace this reality will not only retain talent but also create happier, more engaged high-performance teams.

As leaders, we need to listen to our employees, understand their needs, and create environments where they can succeed both professionally and personally. This is how we build the future of work: by fostering workplaces where flexibility and empathy aren't just buzzwords, but foundational values rooted in the core of the company's ethos.

HARMONY OF WORK, LIFE, AND FAMILY IS TODAY'S TETHER.

With the 'why' of human-centric leadership established in Section 1, we now turn to the 'how.' Section 2 introduces the 26 traits of the **MORE Leadership Framework** – the practical disciplines that will help you and your team **Become More**.

BLANK – FOR NOTES

SECTION 2

THE MORE LEADERSHIP FRAMEWORK

The **MORE Leadership Framework** is the discipline: 26 traits across four pillars — Mindfulness, Opportunity, Relationships, and Execution — practiced with daily intent to thrive, grow, and lead with impact.

Section Two contains the chapters for each of the 26 traits, each one a guide to developing the human qualities that set leaders apart in an AI-first world.

M | MINDFULNESS
Mindfulness, Self-Awareness, Emotional Intelligence, Gratitude, Resilience, Intuition, Balance

O | OPPORTUNITY
Optimism, Hope, Perseverance, Generosity, Motivation, Curiosity, Vision

R | RELATIONSHIPS
Empathy, Social Awareness, Listening, Conflict Resolution, Sincerity, Kindness

E | EXECUTION
Courage, Critical Thinking, Decision-Making, Creativity, Adaptability, Temperance

BLANK – FOR NOTES

M | MINDFULNESS

Mindfulness is where the **MORE Leadership Framework** begins. In a world that pulls our attention in every direction, presence has become a rare skill and a competitive advantage.

To be mindful is not to escape the pace of life but to stand steady inside it. It is noticing. It is choosing. It is returning. It is the discipline of attention in an economy that constantly fractures it.

Leaders who practice mindfulness sharpen emotional intelligence, create space for reflection, and show up fully for the moments that matter most. Presence is the foundation because without it, nothing else holds. You cannot lead with courage, empathy, or vision if you are absent from the moment in front of you.

In this pillar, we will explore seven traits that strengthen presence: Mindfulness, Self-Awareness, Emotional Intelligence, Gratitude, Resilience, Intuition, and Balance.

Becoming More begins here: learning to be here, now.

M | MINDFULNESS
Mindfulness, Self-Awareness, Emotional Intelligence, Gratitude, Resilience, Intuition, Balance

BLANK – FOR NOTES

10 | MINDFULNESS – STAYING PRESENT IN AN ALWAYS-ON WORLD

Your Well-being is What Matters

Mindfulness sharpens presence amid distraction.

Emotional intelligence is a cornerstone of leadership and growth. Mindfulness is one of the most powerful ways to cultivate it. Mindfulness sharpens our ability to stay present and focused, even amid life's many distractions or, for some, a surprising lack of them.

Why Mindfulness Matters

In today's hyper-digital world, distractions are everywhere. On video calls, it's easy to get pulled into in-call chats, message notifications constantly dinging, project management comment notifications, or email alerts (and of course phones). Working from home adds its own layer of interruptions: pets barking, packages arriving, or shared spaces with family members breaking your flow.

On the flip side, for those who work and live alone, the lack of distraction can lead to feelings of isolation or over-focusing on work. While some may crave fewer interruptions, others may need to intentionally break up their day with human connection or a change in pace to stay mindful and energized. Mindfulness teaches us to navigate this dichotomy: avoiding distractions that pull us away when our full attention is needed, while also recognizing when to create healthy breaks to decompress, clear the mind, and return refreshed.

That's why leading organizations like Google and SAP have rolled out mindfulness training, not as wellness perks, but because present, focused employees make better decisions and collaborate more effectively.

What is Mindfulness?

At its core, mindfulness is the practice of being fully present and aware of what you're doing, feeling, and experiencing, without being overly reactive or overwhelmed. It's not just meditation (though meditation can help). Mindfulness can be practiced in everyday moments, like listening carefully during a meeting, stepping outside for fresh air, or pausing before reacting to a distraction.

Five Strategies to Build Mindfulness

1. **Find nearby favorite outdoor places**: Identify a park, trail, or quiet outdoor space close to your home or work. If you don't already have one, **the task is now to find one, or several**. The act of exploring and discovering these spaces is itself an opportunity to practice mindfulness, as it allows you to connect with your surroundings and be fully present. Once you've found your place, spending even a few minutes there regularly can help clear your mind, reduce stress, and bring you back to the present.

2. **Single-task during meetings**: When on video calls or in virtual discussions, commit yourself to focusing all of your attention on the meeting. If you need to, turn off your messaging platform, close your email, and silence other notifications. Commit to fostering connection, learning, and contributing. Avoid fragmented focus.

3. **Use micro-mindfulness techniques**: Incorporate quick practices into your day, such as sixty seconds of focused breathing or a two-minute body release focused on relieving tension. These small moments can have a big impact.

4. **Pause before reacting**: When an interruption happens, a barking pet, a delivery, or an unexpected task, take a deep breath before responding (if at all). This pause creates space to refocus instead of getting pulled off course.

5. **Create intentional connection or distraction**: If you work and live alone, build moments of connection into your day. Call a friend, visit a neighbor, or take a walk to break up your routine. Purposeful distractions can re-energize you and support mindfulness by adding variety to your day.

A Personal Perspective on Mindfulness and Photography

I have several favorite outdoor spots near my home and work that I turn to when I need a mindful reset. On particularly demanding days when I can't visit them, I make time for a mental minute or two to close my eyes and visualize myself in one of those spaces. It's a simple way to center myself and look forward to the next opportunity to be there.

Photography is another tool I use. Capturing picturesque locations allows me to pull up a photo when I need to pause, reflect, and relax. A friend of mine shared how he also uses **photography as a form of meditation**, a way to slow down and be present. It's inspiring to see how mindfulness can take many forms. You can check out some of my favorite places on my Instagram[1]. These small, intentional moments help keep me grounded and reduce stress, even on the busiest days.

How to Become More

Mindfulness isn't perfection; it's choosing presence on purpose. Use these prompts to ground yourself in presence and reclaim focus in a distracted world.
- When was the last time you felt fully present? What were you doing?
- What are your biggest distractions each day? Commit to removing one of them for just 30 minutes tomorrow.
- If you work and live alone, what's one small habit you can add to bring more connection or variety to your day?

One MORE Thing

Mindfulness is self-awareness and reminder of what only humans can do - be mindful, be present. Each moment of mindfulness sharpens your focus, strengthens your leadership, and steadies the path to **Becoming More**.

11 | EYES UP. PHONE DOWN. A REFLECTION ON PRESENCE

How Self-awareness Helps Us Savor the Moments That Matter Most.

The Gift of Self-Awareness

When our whole family is back under one roof, I'm reminded to slow down and savor the now. No matter how challenging life gets, those moments ground me. They prompt me to pause and reflect, take stock of what I have, and find gratitude, even in the hardest seasons.

Self-awareness is more than knowing yourself; it's recognizing how your focus, presence, and actions — or lack of them — impact the people and moments that matter most. Self-awareness invites us to consider an important question: How aware are we of when we're truly present with others? And when we're not, what can we do to improve?

A Personal Reflection on Family and Presence

Time moves so quickly, and life pulls us all in different directions, but being together again, after long separations, reminds me to slow down and fully embrace the joy of being with the people I love most.

I've carried a deep appreciation for these reunions since I lost my grandfather at twelve to a sudden heart attack. In an instant he was gone from the sidelines of my baseball games, my car rides to school, my life. That loss taught me how fragile life is and how fleeting our moments together can be. So when the house is full, it's more than a family gathering. It is a deliberate practice of savoring the now.

The Challenge of Being Present in a Distracted World

In today's fast-paced, always-connected world, it's harder than ever to be present. Studies show the average person checks their phone more than 90 times a day. We're pulled in so many directions, distracted by the constant noise of notifications, emails, and numerous different feeds. To combat the myriad of pulls on our attention, I offer a simple mantra:
Eyes up. Phone down.

Try to look up more than you look down. Put the phone aside. Not to abandon it entirely, but to create space for real human connection. Snap the quick photo, answer the occasional text, but then return to the room, to the people, to the moment.

> Presence starts with small choices, like simply lifting our eyes and focusing on what is right in front of us.

The same holds true at work. I've seen leaders transform team trust simply by closing their laptops in meetings and giving people their full attention. Presence communicates value and respect. Nothing erodes trust faster than someone feeling like you're half-listening while checking email and sending messages.

Practicing Presence: A Self-Awareness Challenge

Self-awareness is noticing when we've drifted and gently bringing ourselves back to the present. It's being mindful of how we're engaging with others and reflecting on how we can be more intentional about truly showing up.

Practice presence, not just as a gift to those around us, but as a gift to ourselves. Start small. What's one habit you can develop to better recognize, being more self-aware, when you're not present? Maybe it's noticing when you reach for your phone during a conversation or catching yourself when your mind starts to wander. With self-awareness comes the power to shift and the opportunity to be more fully present.

How to Become More

Self-awareness begins with noticing. Use these prompts to examine how you show up for others and yourself.

- When was the last time you caught yourself drifting away from the moment? What cue could you use to pull yourself back faster?
- What habits most often pull you out of presence? And how can you return with intention?
- Who in your life deserves more of your undivided attention? Schedule time with them this week — and protect it from distraction.

One MORE Thing

Self-awareness begins with acknowledging that our presence, the awareness of our surroundings and others, is an active human behavior, complicated by the attention and distractions of an AI-first, always-on world. Notice when you drift, return with intention, and offer your full self to the people and moments that matter most.

TRAIT: EMOTIONAL INTELLIGENCE

12 | EMOTIONAL INTELLIGENCE IN MODERN LEADERSHIP

The Transformative Role of Emotional Intelligence in Leadership

While we often hear about strategic thinking, decision-making, and execution as hallmarks of great leadership, emotional intelligence is quickly becoming an equally essential trait. One that allows leaders to connect with their teams, understand others, navigate complexity, and foster cultures of trust and resilience.

Emotional Intelligence: The New Cornerstone of Leadership

In today's world, dominated by rapid change and AI-driven transformation, the ability to understand and manage emotions has never been more relevant. Emotional intelligence is about more than being empathetic or "nice." It's about using self-awareness and understanding others' emotions to build stronger relationships and inspire action.

Some of the most admired leaders in the world embody emotional intelligence, even if they don't always name it explicitly. For example, Jacinda Ardern, former Prime Minister of New Zealand, said,

> *One of the criticisms I've faced over the years is that I'm not aggressive enough or assertive enough, or maybe somehow, because I'm empathetic, it means I'm weak. I totally rebel against that. I refuse to believe that you cannot be both compassionate and strong.*

Her leadership through crises, such as the Christchurch mosque shootings demonstrates the power of balancing empathy with decisive action, a hallmark of emotional intelligence. This balance, often seen in transformational leaders, is equally critical in corporate environments, where leaders must navigate high-stakes decisions while fostering connection and trust. Indra Nooyi, former CEO of PepsiCo, underscores another key aspect: caring.

> *As a leader, I am tough on myself and I raise the standard for everybody; however, I am very caring because I want people to excel at what they are doing so that they can aspire to be me in the future.*

Caring with intent fosters collaboration and trust, two outcomes that are increasingly critical in today's hybrid, remote workplaces.

I've also seen the absence of emotional intelligence play out in my own leadership. After a company reduction, I was tasked with announcing a salary percentage increase to the broader team. My focus was too much on the positive aspect — that an increase was being given at all — and I failed to take full stock of the team's emotional state after losing colleagues. I delivered the update too coldly, missing an opportunity to establish trust and demonstrate empathy. The lesson was clear: before you speak, put yourself in your audience's shoes. Ask: *"How would I want to receive this news?"*

Transactional vs. Transformational Leadership

The concepts of transactional and transformational leadership were first introduced by political scientist and historian James MacGregor Burns in his seminal 1978 work, *Leadership*[1]. Burns argued that leaders could operate on two ends of a spectrum: **transactional leaders** focus on exchanges of value (tasks for rewards) while **transformational leaders** seek to inspire and elevate both their followers and themselves.

These two styles offer a powerful lens through which to view emotional intelligence in action. Transformational leadership aligns closely with emotional intelligence. It relies on empathy, deep understanding, and a shared vision to motivate and connect with others. Further descriptions:

Transactional Leadership: Focuses on routine, established procedures, and rewards or punishments based on performance. While effective in stable environments, it may lack the flexibility and innovation needed in today's dynamic hybrid and remote workplaces. Transactional leadership, while less emotionally dynamic, remains valuable in environments where consistency, compliance, and operational efficiency are key.

Transformational Leadership: Inspires and motivates by creating a vision, fostering creativity, and encouraging personal development (helping others **Become More**). This style aligns closely with emotional intelligence, as it requires understanding and addressing the needs and emotions of team members.

Research[2] indicates that transformational leadership is often associated with higher levels of emotional intelligence. Leaders who are emotionally intelligent are more likely to engage in transformational behaviors, such as inspiring and intellectually stimulating their teams. Conversely, transactional leadership, with its focus on structure and control, may not leverage emotional intelligence to the same extent.

Why Emotional Intelligence Matters

Emotional intelligence is not yet universally recognized as a leadership must-have, but it's becoming clear that leaders who lack it are falling behind. Consider these ways emotional intelligence can elevate modern leadership:

1. **Adaptability in Change**: Emotional intelligence helps leaders manage their own reactions to uncertainty while guiding their teams through transitions.

2. **Building Psychological Safety**: By fostering trust and open communication, emotionally intelligent leaders create environments where people feel safe to take risks, admit failure, and innovate.

3. **Conflict Resolution**: Leaders who understand emotions can turn disagreements into opportunities for connection and growth.

4. **Supporting Well-Being**: Leadership today involves helping individuals and teams navigate not only work stress and complexity, but also home stress and complexity (such as childcare or eldercare), while maintaining performance and morale.

How to Become More

Emotional intelligence is developed, not inherited. Use these prompts to build your ability to connect, empathize, and lead with trust.

- How well do you understand your own emotional triggers?
- Do you create space for others to share emotions, ideas, or fears without judgment?
- Think of a recent message you delivered to your team. Did you frame it from your perspective, or from theirs? If you had to put yourself in their shoes, how would you change the way you delivered it?

One MORE Thing

Emotional intelligence grows with practice. Each choice to reflect and connect expands your capacity and reminds us that empathy, care, and trust are traits no machine can truly replicate.

TRAIT: GRATITUDE

13 | GRATITUDE IN ACTION: RECOGNITION AS A DAILY PRACTICE

Seeing Courage, Families, and Everyday Leadership Through the Lens of Gratitude

Gratitude is the discipline that sharpens presence into mastery. When we pause to honor those who serve, we're practicing more than courtesy, we're learning to hold another person's sacrifice in full view, uninterrupted by our own noise. That act rewires our attention toward what's noble in the everyday, turning recognition into a habit that elevates teams, families, and communities.

In the **MORE Leadership Framework**, Gratitude lies within the pillar of **M: Mindfulness** because it trains the leader's eye to see value before seeking value. Every day **we should HONOR,** not only support, those who serve, and in doing so, strengthen the reflex to say, "I see you, I appreciate you," wherever we lead next.

In leadership, gratitude isn't just courtesy — it's strategy. Research shows that employees who feel genuinely appreciated are more engaged, more productive, and more resilient under pressure. Gratitude creates teams that can weather disruption because people feel seen and valued.

Reflecting on Courage

When we honor those who serve, the chapters on kindness and courage take on even deeper significance. Veterans embody these traits profoundly. They demonstrate not only the courage to face physical and mental challenges but also the kindness and commitment to protect and serve a community beyond themselves. Their courage to serve, knowing the sacrifices it may demand, parallels the quiet courage we should cultivate in our own lives, standing firm in our values, examining our flaws, and contributing to the greater good.

Honoring the Sacrifices of Families

Service extends beyond the individual. The courage is evident in the uniform, but just as significant are the ripple effects. I'd like us to reflect on a few often-overlooked aspects of the sacrifices made for our country. Their courage is evident in the uniform, but just as significant are the ripple effects of that service, the quiet, enduring sacrifices made by the families who stand by them.

Every deployment leaves an empty place at home.

Spouses, parents, and especially children carry the weight of separation, navigating life with an ever-present hope for their loved one's safe return. These families are steadfast in ways many of us may never fully understand. Waking each day with wonder and uncertainty, powerless over when they may hear from their loved one next. They embody resilience and strength that is difficult to put into words and often goes without the recognition their sacrifices deserve. These families deserve not only our gratitude, but also our support.

Gratitude begins with a selfless mindset — noticing and naming the value in others. Kindness, which we'll explore later, is gratitude in motion: proactive help, empathy, and generosity. By cultivating gratitude first, leaders fuel the very traits that translate outward into action.

The Hidden Struggles of Veterans

Beyond the visible courage of service members and the sacrifices of their families lie struggles we too often overlook or fail to understand. Post-traumatic stress disorder (PTSD[1]), the scars of combat (including conditions such as Moral Injury[2] and Operator Syndrome[3]), both visible and invisible, and the lasting impact on physical and mental health are realities many veterans carry long after returning home.

These challenges don't end when deployment does; they are often woven silently into daily routines, families, and workplaces. For many, the workplace becomes a space where they must navigate the balance between extraordinary strengths and unseen burdens. Recognizing this requires us to move beyond surface gratitude, offered on just one day of the year. To truly honor them, we must cultivate genuine understanding, support, and inclusion every day.

Supporting Veterans in the Workplace

Then there's the ongoing journey of those who come home with visible and invisible scars. Many veterans of the Gulf Wars are now part of our workplaces, and among them are individuals navigating post-traumatic stress disorder (PTSD). Having had the privilege of working alongside someone affected by PTSD from their time in service, I understand that the impact of deployment extends well beyond the battlefield. These men and women bring extraordinary strengths, yet they also carry profound burdens that may be invisible to those around them.

I encourage leaders to acknowledge this issue in the workplace. If a veteran employee is open to it, begin an open, supportive dialogue to express your willingness to support them. We must remember that while it seems there is a barrier between our work lives and our home lives, to help an individual discover their full potential, this barrier needs to be removed, and we must help the whole person **Become More**.

Addressing the Veteran Suicide Crisis

Finally, we must acknowledge the heartbreaking crisis of veteran suicide. Reports consistently highlight a persistently high rate. While there is some debate over the exact daily number, that debate seems trivial in comparison to the fact **that we are losing any veterans at all to suicide**. This issue speaks to the weight many veterans bear alone. A weight we cannot ignore. A weight that demands not only awareness, but action in being more supportive of those who have served, who are serving, and will serve in the future.

Moving Toward Deeper Commitment

Gratitude deepens when it becomes action. To honor service is to recognize the courage it takes to serve, the quiet strength of the families left waiting, and the unseen struggles that many face long after their service ends. In doing so, we move beyond simple gratitude, toward a more meaningful commitment to those who have given so much.

Gratitude also acts as a buffer against stress. Teams that practice it regularly are more adaptable, more connected, and better able to face uncertainty together. In an AI-first, always-on workplace, this resilience may be one of the most important outcomes of gratitude in action.

How to Become More

Gratitude is more than words. It's recognition in action. Use these prompts to turn gratitude into a daily discipline.
- Where in your life can gratitude become action, not just words?
- Who in your circle deserves recognition that you've never voiced?
- How might practicing daily gratitude reshape the way you see ordinary moments?

One MORE Thing

Gratitude is leadership in its simplest form: noticing, naming, and honoring others. Each time you act on it, you multiply trust, resilience, and connection.

14 | THE ART OF RESILIENCE: STRENGTH DURING UNCERTAINTY

Building the Mindset to Navigate Change with Confidence and Purpose

My upbringing shaped my perspective on perseverance, growth, and the pursuit of more. **Resilience** is essential to progress.

The Weight of Loss and the Strength to Continue

Resilience is often framed as the ability to bounce back from setbacks, but there's a deeper, more human side to it. Sometimes resilience isn't about bouncing back quickly. It's sitting with the weight of what happened, acknowledging pain, learning from it, and then moving forward with intention.

When tragedies strike, whether close to home or across the world, we're reminded how fragile life is. In moments like these, the demands of work can feel trivial. As leaders, our role is not to rush forward as if nothing happened, but to pause, acknowledge the weight others are carrying, and offer presence and support. True resilience isn't simply how we recover individually; it includes how we help others find the strength to keep going when life gets hard and feels overwhelming.

And resilience isn't about gritting your teeth and pushing forward no matter the cost. Sometimes the most resilient choice is to pause, to heal, or to ask for help. That kind of strength prevents burnout and reminds us that vulnerability is not weakness but part of enduring and recovering well. Loss, however, is not always sudden or tragic. Sometimes, it comes in the form of saying goodbye to an elderly loved one. These moments carry their own weight, not of shock, but of reflection, of honoring a life well lived and finding gratitude amidst grief. Resilience in these moments is about remembering, celebrating, and carrying forward their lessons.

Real resilience is not being unaffected by loss. It's having the strength to pause, to care, and to lead with empathy before helping others find their way forward. Our first responsibility is always to the person and what they need as a human, not what the business needs from an employee.

Resilience and the Future of Work

Work and life are no longer separate. The idea that what happens at home stays at home is outdated, what affects us personally follows us into our work. Leadership today requires understanding that resilience at work starts with acknowledging the full human experience. When an employee faces loss, stress, or personal challenges, it doesn't get left at the door when they log in or step into a meeting.

Resilient organizations recognize that their people are more than their output. They create cultures where employees feel supported, not just in professional challenges but in personal ones as well. Resilience is deeply personal. Leaders instill it in their teams by modeling calm in crisis, creating space for recovery, and showing that setbacks are not the end but part of the journey. A resilient culture emerges when people see that it's safe to stumble, regroup, and try again.

This is the new standard for leadership: seeing the whole person, fostering an environment where life's realities are acknowledged, and ensuring that support structures exist to help people move forward.

How to Build Resilience in the Face of Setbacks

Whether we're facing personal hardships, professional failures, or collective tragedies, resilience is what allows us to *honor what's been lost* while continuing forward. Here's how we can cultivate it:

- Acknowledge Reality, Don't Avoid It.
  - o Pain and setbacks are real, and pushing past them too quickly can be harmful. Allow space for reflection and healing. The passing of a loved one, particularly an elderly relative who has lived a full life, can bring both grief and gratitude. Allow space to reflect on their impact while finding comfort in their legacy.
- Connect Before Correcting.
  - o Before trying to fix, solve, or move forward, connect with those affected. Be present. Offer support. In the workplace, that might mean simply acknowledging that things feel heavy before jumping into the next project.
- Reframe the Narrative.
  - o Resilient people don't ignore pain, but they *do* work to find meaning. What can be learned from this challenge? How do we honor loss while building forward? "*The unexamined life is not worth living.*"
- Lean on Others.
  - o Resilience isn't a solo act. Whether through community, mentorship, or family, surrounding ourselves with support helps us navigate the hardest moments.
- Focus on What's Within Your Control.
  - o In moments of tragedy, some things are simply out of our hands. But we can always control how we show up for others, how we listen, how we express kindness, and how we choose to move forward.

Leadership in Times of Loss

As leaders, resilience doesn't push people past their emotions. It holds and allows space for grief, uncertainty, and recovery. When tragedy or loss strikes, check in with your people. Whether it's an unexpected event or the passing of a beloved elder, acknowledging their grief and offering support is more important than pushing for productivity. I remember one manager who, after a teammate lost a parent, said simply: "Take the time you need. We'll cover for you. When you're ready, we'll be here." That short exchange did more to build loyalty and trust than any project success could. Resilience isn't built by pretending work is untouched by life. It's built by showing up human first and doing what's best for the individual in the moment.

How to Become More

Resilience doesn't ignore pain. It endures with clarity and care. Use these prompts to reflect on how you recover and how you support others in times of loss or challenge.

- How do you personally process setbacks or difficult events?
- What's one way you can show up with resilience and empathy for someone in your life right now?
- Where in your work or relationships could you create more space for others to recover, before pushing for progress?

Later we'll explore perseverance, the long drive to finish. Resilience is different. It's the emotional endurance that steadies you in the hardest moments, so you can rise and keep moving.

One MORE Thing

Resilience balances grit with grace. It asks us to face difficulty honestly, to support others through it, and to keep moving forward with intention.

TRAIT: INTUITION

15 | THE COMPASS WITHIN: INTUITION AS A LEADERSHIP EDGE

Building Everyday Instincts That Cut Through Noise and Doubt

Intuition is often described as a gut feeling; an inner knowing that guides us when logic alone falls short. For many leaders, it's more than instinct. It becomes an edge. Intuition sharpens judgment in ambiguous moments, helps sense what data can't yet prove, and enables decisions that build trust, culture, and momentum. In leadership and personal growth, learning to recognize and rely on intuition is key to navigating complexity with speed and confidence.

Intuition is a critical leadership skill because it guides us when facts run out or when multiple paths seem equally viable. Paired with critical thinking, it becomes a powerful edge, allowing leaders to act with both speed and sound judgment. It's the tug that tells you to slow down when everyone else is charging ahead. Intuitive leaders aren't distracted by the crowd — they listen, they assess, and they pause and reflect.

Intuition at Work: Gauging Contribution

The place where intuition most often shows up for me is in gauging contribution and productivity. Within the principles of Modern Work Ethics, contribution is about showing up and producing work that matters. But with remote knowledge workers, it isn't always easy to measure the value of forty hours behind a laptop. Metrics and dashboards can tell one story, but sometimes they don't capture the full picture. Over the past few years, I've learned to pay attention when something doesn't feel right. If my intuition signals that an individual's contribution is below a reasonable expectation, I don't ignore it. I dig in. Left unchecked, a lack of contribution is often visible to peers, creating disruption and even resentment within the team.

In one case, my gut told me the output from a team member wasn't aligning with what the role required. Rather than jump to conclusions, I asked for a simple daily report of work being completed. Within days, the gaps became clear. Data eventually confirmed what my intuition had already sensed; contribution wasn't where it needed to be. Acting on that instinct allowed me to reset expectations, support improvement, and uphold the principle that contribution matters.

Assess Yourself: Intuition

Where do you currently stand in terms of intuition? How often do you trust your instincts? Use these questions to help assess your current relationship with intuition:

- Is intuition a trait you've considered in relation to your success at work? Have you considered how your intuition differentiates you from others?
- Do you frequently rely on your intuition when making decisions? If not, why not?
- In what areas of life (personal or professional) do you tend to trust your gut feelings? Where do you ignore them?
- Are you more likely to second-guess yourself, or do you confidently follow your instincts? When you second-guess yourself, why?
- Has following your intuition led to any poor or hasty decisions?

Use these questions as a baseline. Then, over a week or a month, track moments when your intuition guided you. Afterwards, return to the reflective prompts to evaluate how your instincts influenced your choices.

Reflective Prompt:

Take a moment before making any significant decision and ask yourself:

"What is my intuition telling me?"

Pause and reflect. This isn't rushing to judgment but giving space to those initial feelings, ideas, and insights. Track them as you move toward your final decision.

After a period of practice, such as a week or a month, take a few moments to reflect on how often you listened to your intuition. Were there times when you wished you had trusted it more? What difference did it make in the decisions where you relied on it? This reflection will help you tune into your instincts more readily going forward.

Considerations For Strengthening Your Intuition:

- **Trust Your Gut**: Start by listening to your first instincts, even if they seem unclear at first. Instead of dismissing them, reflect on where those feelings might come from.
- **Practice Quieting Your Mind**: Take moments throughout the day to quiet your thoughts. Meditation, deep breathing, or taking walks can help you be more attuned to subtle insights.
- **Reflect on Past Decisions**: Look back at times when you followed your intuition. How did it turn out? Were there moments when you ignored your gut and later regretted it? Use these reflections to reinforce your connection with your inner guidance.
- **Create Space for Reflection**: Intuition often emerges when you give yourself the space to process. Build time into your routine to step back and reflect on decisions instead of rushing through them.

How to Become More

Intuition is your inner compass. It sharpens with practice, trust, and reflection. Use these prompts to reflect on how you listen to your instincts and where they guide you.

- When making decisions, do you pause enough to sense what your intuition is telling you?
- Where in your life have you ignored your gut and regretted it later?
- What small habit could you build to create more space for intuition to surface?

One MORE Thing

Intuition is your compass. It sharpens with practice, reflection, and the courage to act before certainty arrives. Each time you listen inward, you strengthen your ability to decide with clarity, confidence, and conviction.

16 | BALANCE: THE ANCHOR TRAIT OF MINDFULNESS

Finding Harmony, Presence, and Perspective in Modern Leadership

Mindfulness, as one trait in the framework and the first pillar, calls us to pause, be fully present, and live with intention; both personally and professionally. When we bring that same presence to balance, keeping work, personal life, and priorities in harmony, it becomes the anchor that steadies everything else. It puts the power to control your time back in your hands. Balance, as the capstone trait of Mindfulness, is what sustains every other discipline in this pillar. Without it, presence becomes fleeting and growth becomes unsustainable.

Balance as Harmony

Balance looks different today than it did even a few years ago. The rise of remote and hybrid work has blurred the lines between our professional and personal lives. With so much work now at home, home often finds its way into work. The concept of "leaving work at work" is impossible now that offices might also be a kitchen table, a bedroom, or a backyard porch.

In this reality, maintaining balance means being mindful of how we spend our time and energy. One must recognize when work begins to creep into family dinner or when personal distractions take over productive hours.

Balance doesn't mean strict separation. Blending has become the norm. It means setting boundaries and creating intentional moments for both work and life.

Personally, I've been working on spending less of my mental energy thinking about work when I'm not at work. I tend to lean in hard, letting my imagination run into strategies, ideas, and to-dos, even during personal time. Recently, with the encouragement of my wife, I've started setting boundaries, like no team-chat after 8:00 and trying to carve out time to fully disconnect. It's been a struggle as creativity in thought is a peaceful place for me, though it pulls my attention from being present with my family. Balance offers permission to make an assessment of life's priorities. Balance is not a one-time achievement but a continuous tuning process, regularly checking in to confirm you're not leaning too heavily one way or the other.

I've learned the hard way that when balance is lost, the result is often burnout, mistakes, and even frayed relationships. Leaders who ignore balance eventually run out of energy and perspective. Leaders who protect it succeed longer, because they last longer.

Balance in Leadership: Positivity, Curiosity, and Intentionality

Balance also extends into how we approach performance and feedback. Leadership requires us to walk a fine line between positivity and scrutiny. It's important to recognize and celebrate wins, ideally following the four-to-one rule, where we offer four pieces of praise for every one piece of constructive feedback. Too often, this ratio is flipped, with criticism far outweighing recognition.

Leaders with high emotional intelligence understand the power of leaning into positivity to motivate and build trust.

That said, balance here doesn't mean avoiding tough conversations or overlooking underperformance. Sometimes, it's necessary to raise intensity and focus to address challenges. However, effective leaders balance critique with curiosity. When someone underperforms, the first instinct shouldn't be to blame. It should be understanding. What's behind the issue? Is it a lack of clarity, support, or resources? Asking thoughtful questions like, *What obstacles are you facing? How can I help you succeed?* shifts the focus from fault to growth.

Another key aspect of balance in leadership is practicing intentionality in how and when we communicate. As leaders, we often spend much of our day thinking about work, and it's easy to default to sharing every thought, idea, or task - the moment it comes to mind. But that constant stream of communication can disrupt the balance your team is working hard to maintain.

I've been working on practicing restraint in this area: jotting down ideas and scheduling emails or messages for later, rather than sending them immediately. Just because I'm thinking about work late in the evening doesn't mean my team needs to be.

This small act of mindfulness helps create a culture where urgency is reserved for what's truly urgent, and employees feel empowered to maintain boundaries without fear of judgment.

Here's what balance can look like in practice: a leader chooses to delegate a late-day task rather than stay two extra hours to finish it. The work still gets done, a team member gains ownership, and the leader gets home in time for family dinner. Balance here doesn't reduce productivity, it enhances it.

By being intentional with communication, leaders model the balance they want to see in their teams. One shows that while work is important, it doesn't need to consume every waking moment. This balance fosters trust, reduces stress, and creates an environment where people can thrive.

Balance in Health and Wellness

Balance also extends to our health and wellness. I've learned that living healthily isn't about being perfect all the time. For years, I thought it meant constant discipline: no dessert, no indulgences, just strict adherence to a plan. But in recent years, I've embraced a more sustainable mindset. It's okay to enjoy a treat without guilt or to skip a workout if you need rest. I follow an 80-20 approach, making good choices most of the time, while giving myself grace when I don't. This perspective has allowed me to stay strong, healthy, and happy without the stress of chasing perfection.

The same is true in leadership. Balance isn't rigidity; it's sustainability. Leaders who demand perfection eventually crack under the pressure, and so do their teams. A balanced leader, by contrast, knows when to push hard and when to pause, when to give feedback and when to offer encouragement. They create an environment where performance is fueled by energy, not exhaustion. Balance, in health and in leadership, isn't about doing everything—it's about doing the right things with consistency, perspective, and care.

Why Balance Matters for Growth

Balance gives us the energy to stay the course and grow sustainably. It's what allows progress to feel steady rather than frantic, intentional rather than reactive. With balance, we conserve energy for the moments that truly matter. It helps us be more present in our relationships, make better decisions, and avoid burnout. Without it, even the best intentions can lead to exhaustion, mistakes, or a quiet loss of joy.

Earlier we explored balance as a societal shift in how work and life intersect. As a trait in the **MORE Leadership Framework**, balance shows up in the personal choices and boundaries we set daily. Balance is a daily practice, intended to make one more productive and creative, not less committed. Balance is what ensures the other traits in the Mindfulness pillar — presence, self-awareness, emotional intelligence, gratitude, and resilience actually last. Without balance, they all become temporary. With it, they endure.

How to Become More

Balance doesn't mean rigid separation. It means intention. Use these prompts to reflect on where your energy goes and how you might create more harmony between work, life, and leadership.
- Where do you feel harmony in your life, and where do you feel imbalance?
- How can you realign your priorities to honor both your professional and personal values this week?
- What small adjustments can you make today to cultivate balance?

One MORE Thing

Balance steadies us. It calls us to civility, empathy, and perspective. It reminds us that true progress is built on harmony, not division.

BLANK – FOR NOTES

M | MINDFULNESS

You get one life.

How you spend your time is all you have.

— Jeremy Victor

Mindfulness roots our presence and gives us power over the distractions fighting for our attention. Its purpose in the framework is to remind us that time is limited and there is one person in charge of the choice of how we spend it. With yours, be mindful.

M | MINDFULNESS

Mindfulness, Self-Awareness, Emotional Intelligence, Gratitude, Resilience, Intuition, Balance

BLANK – FOR NOTES

O | OPPORTUNITY

Optimism and hope are not luxuries in leadership. They are necessities yet often are neglected or misunderstood. The ability to see possibility where others see limits is what allows teams, organizations, and communities to move forward.

Opportunity begins with a choice: to believe that the future is not fixed. That setbacks can be turned into setups. That obstacles can become openings. That struggle can be rewarding.

But possibility is not passive. It requires leaders who are both hopeful and grounded, who can inspire belief while also preparing for reality. In a complex, always-on world, we need more than problem-solvers. We need leaders who generate possibility, who create energy and momentum when others stall.

In this pillar, we will explore seven traits that shape possibility: Optimism, Hope, Perseverance, Generosity, Motivation, Curiosity, and Vision.

O | OPPORTUNITY:
Optimism, Hope, Perseverance, Generosity, Motivation, Curiosity, Vision.

BLANK – FOR NOTES

17 | OPTIMISM – THE FUEL OF FORWARD

Rethinking Optimism: It's Not Binary. It Can Be Learned, Refined, and Strengthened.

Optimism is often perceived as an inherent trait. Is your glass half full or half empty? That framing is limiting because it implies you're either one or the other, and that optimism can't be developed or strengthened. Optimism doesn't hinge on half-full or half-empty. It rests on what you do with what is in the glass.

My Journey from Cynicism to Optimism

Optimism is a part of me, rooted as the "O" in my nickname, though it wasn't always that way. Entering college, after not getting into my first-choice school, I was pissed off at the world. And I lived my life that way. Hard. I remember buying a "Mean People Suck" sticker, cutting off the word 'Mean,' and sticking "People Suck" on the dash of my 1979 CJ-5. I'll spare you the details of my troubled times but suffice it to say, optimistic I was not.

That cynicism carried into how I treated people. I pushed friends away, assuming they'd let me down anyway. I skipped opportunities, telling myself they wouldn't work out. Even in small things, like dismissing group projects before they began, I was quick to see what was broken instead of what could be built. Looking back, I realize how much energy I wasted protecting myself with anger and cynicism instead of investing in possibility.

It wasn't until I started my first professional job that I fully realized how optimism can be nurtured into a powerful force for forward momentum. After an Operations Department reorganization, I was left disappointed with the position I was given. It was the more junior role in the department, and I felt certain I was capable of the other. At the time, my ego wouldn't let me hear the reasons why. I knew they were wrong. *(The things you wish you could tell your younger self. I digress.)*

Shortly after the reorg, I was out holiday shopping, browsing one of those pop-up kiosks that sell calendars, motivational posters, and plaques. I happened on one with the word **Optimism** and the quote: *"Every obstacle is a stepping stone to your success."* Decades ago, I bought a small, framed quote. I still have it today. I've been overcoming obstacles ever since. But here's something that many people misunderstand: **Optimism alone isn't enough.**

The Problem with Blind Optimism

For a long time, I thought optimism was simply about believing things would work out. That's what we're told, right? *Just stay positive.* But optimism, by itself, **can be dangerous**. Blind optimism ignores risks. It convinces us to charge forward without questioning the path. It tells us everything will be fine, even when the warning signs are flashing red.

History gives us sharp reminders. In 2008, Blockbuster CEO Jim Keyes said, "Neither RedBox nor Netflix are even on the radar screen in terms of competition.[1]" His optimism in the status quo blinded him to the shift already underway. Within two years, Blockbuster filed for bankruptcy while Netflix soared. Leaders who rely only on optimism set their teams up for failure.

Here's the flip side: **skepticism without optimism is just cynicism.** It keeps us stuck, focusing only on problems instead of solutions. The key isn't choosing between optimism or skepticism. It's knowing how to use both.

Grounded Optimism: Belief Meets Critical Thinking

The best leaders aren't just optimistic. They are *"grounded"* **optimists**: balancing belief with realism, vision with execution, and hope with adaptability. And no one embodies this better than **Jürgen Klopp**.

Klopp didn't just turn Liverpool FC into a winning team. He built a *mentality*. He infused his players with belief, convincing them that no game was ever over, no challenge too big. But he didn't stop at belief. He was also brutally realistic. When things weren't working, he adapted. He analyzed weaknesses, changed tactics. His optimism wasn't blind. It was calculated.

The Barcelona Miracle: Optimism in Action

In the 2019 Champions League semifinals, Liverpool lost the first leg **0-3** to Barcelona. No team had ever come back from that kind of deficit. Most managers would have accepted defeat. Not Klopp. Before the second leg, he told his team:

> *"It's impossible. But because it's you, it's possible."*

That wasn't false hope. That was *grounded* **Optimism**. He made them believe. But he also made them prepare. He studied Barcelona's weaknesses, adjusted tactics, and instilled the *relentless* mindset that led to a **4-0 comeback victory**. Klopp used optimism to rally belief in the face of impossible odds. Sara Blakely, founder of Spanx, used optimism to start something no one believed in. Both show that optimism paired with action is how to creates results. That's the balance. Optimism gives people the fuel to push forward. Skepticism makes sure they don't drive off a cliff.

How to Build Optimism Within Your Leadership

So how do you cultivate optimism *without losing sight of reality*?

Believe in Possibilities, but Pressure-Test Reality
- Instead of assuming things *will* work out, ask: What would need to be true for this to succeed?
- Prepare for obstacles ahead with an open mind.

Adopt a *Forward Thinking Readiness* Mindset

- Instead of waiting for failure and analyzing what went wrong (*post-mortem*), ask upfront: If this idea or plan were to fail, why would that happen?
- This isn't negativity. It is intuition and preparation.
- Optimism should never be blind. It should be ready: ready for obstacles, ready for adjustments, ready to succeed because you've already accounted for what could go wrong.

Differentiate Between Vision and Execution

- Be wildly optimistic about *where* you can go.
- Be brutally skeptical about *how* to get there.
- Example: Sara Blakely (Spanx). She had a vision to change the shapewear industry, but she didn't rely on optimism alone. She cold-called manufacturers, tested different materials, and handled distribution herself until she proved the concept. Optimism gave her the courage to start; execution made her successful.

Watch for Confirmation Bias

- Optimists tend to seek evidence that supports their beliefs.
- Skeptics look for data that challenges assumptions.
- The best leaders use critical thinking and consider both before making decisions.

How to Become More

Optimism isn't blind belief. It is belief paired with preparation. Use these prompts to ground your optimism to fuel progress without ignoring reality.

- Where in your life or work do you lean *too far* into optimism or skepticism?
- If you're **too optimistic**, ask: What's blind spots am I missing?
- If you're **too skeptical**, ask: What if this actually worked?

One MORE Thing

Optimism is the fuel of forward. Skepticism keeps ambition grounded. The best leaders know how to use both to accomplish hard things.

18 | HOPE IS A RESPONSIBILITY

What Jackie Robinson Can Teach Us About Hope

I spend a lot of time thinking about what it must have been like to be the first.

Not in the way we celebrate it now, with documentaries and commemorations. But in the actual moment. The day-to-day grind of being the *first to walk a path* no one believed was possible.

Take Jackie Robinson.

Imagine stepping onto the field in 1947, the weight of history on your shoulders. Imagine walking into stadiums where entire crowds weren't just rooting against you. They didn't even think you belonged there. Imagine striking out in front of those people. Taking that long, lonely walk back to the dugout while thousands of voices screamed at you, not just because you missed a pitch, but because they wanted you to fail. And yet, he kept going.

That's hope.

Not just optimism. Not just a wish that things might get better. **Hope is the relentless belief that better will come,** even when the moment feels impossible. Hope is knowing that even when today feels impossible, tomorrow is another chance. I think about this all the time. How easy it is to get caught up in our own struggles, to feel like things are too hard. But then I put myself in the shoes of someone like Jackie Robinson.

And I ask myself: **Is my journey really that hard?**

Hope isn't passive. It's not sitting around waiting for things to improve. It's picking yourself up after failure. It's showing up again tomorrow. **It's the belief that, no matter how bad today looks, good will prevail ... eventually.**

Hope is a responsibility we carry forward in today's world of accelerating technology. In an AI-first, always-on economy, it's not enough to hope we won't be replaced or left behind. Our responsibility is to hope for and build toward a future where human creativity and technology together make work more meaningful, not less. Hope here isn't naïve. It's an active choice to shape progress so it serves people, not just systems.

How to Become More

Hope is not passive. It is the belief that better will come and the willingness to keep moving when today feels impossible. Use these prompts to anchor your own hope.

- What's one challenge in your life that feels overwhelming?
- Are you carrying hope through it, or just waiting for things to change?
- How can you show up today, even when it's hard, believing that better is ahead?

One MORE Thing

Hope isn't only believing in the future, it's fighting for it. Hope isn't passive. It is the courage to keep going. Step up, take your swing, and believe the next attempt might be the one that changes everything.

TRAIT: PERSEVERANCE

19 | THE WISDOM TO PERSEVERE AND THE WISDOM TO WALK AWAY

Perseverance is Powerful. Applied Blindly It May Lead You in the Wrong Direction.

What taught me perseverance?

Batting slumps.

I played a lot of baseball growing up. Senior season in high school ... the worst time for a slump for an aspiring college athlete ... It was so bad that after the 7th or 8th game I was benched. I went from tying for the batting title as a junior to the bench. I didn't actually break out of that slump until summer league. At the time, I thought it was just a slump. But later, I understood that my struggles at the plate weren't just about baseball. They were about something much bigger.

Baseball taught me perseverance is the discipline to keep showing up, adjusting, and grinding through discomfort. It also taught me that failure is feedback. Baseball gave me the first taste of what it means to persist toward a goal even when momentum has turned against you.

Fishing.

I was lucky enough to grow up with a pond within walking distance of my house. I spent hours of casting and waiting, often walking home empty-handed. But the thrill of reeling in a largemouth bass kept me going back.

Little did I know at the time that fishing not only taught me patience, but also that perseverance is often less about one cast and more about hundreds. It's the same lesson leaders need in long, uncertain projects to keep showing up, knowing results rarely come instantly.

Baseball gave me perseverance in the moment, fishing gave me perseverance over the long run. Together they taught me that persistence looks different depending on the challenge.

My birthday, starting at ten.

My 10th birthday was the first without a card from my dad after he left. It would be 34 years before I received another. Those lonely years as a kid and teenager taught me resilience and perseverance in the face of an obstacle that amongst the people I knew, I was the only one going through. Learning early that expectations and reality don't always match, and that disappointment doesn't stop the world from moving forward.

This is where resilience and perseverance overlap. Yet, there is an important distinction in the **MORE Leadership Framework**. Resilience steadied me in the loss itself; perseverance carried me through the years that followed. Resilience helps us recover; perseverance sustains us toward the goal still in front of us; motivation is the inner drive that pushes both forward.

Perseverance is persistence despite difficulty or delay in success. It's like a muscle, the more you have to endure, the stronger your perseverance, the stronger your drive to move beyond the obstacles. Adversity. The randomness of life. It seems like the human psyche is wired for battle, always struggling against something, whether external or internal. But here's the irony:

Perseverance isn't always the right answer.

It has to be checked against purpose. Goals that once mattered can lose relevance, and without reflection, effort turns into wasted energy. I once had a working relationship that started off rocky. Early on, I was making the other person feel like I didn't value their contributions. So I made the effort. I gave more to the relationship than I was getting in return, believing that with enough time and persistence, things would shift. And in some ways, they did. We worked together. We functioned.

But after years of trying, I came to an important realization: not every relationship has to be a great one. And if you're putting more into a relationship than the other person, at some point, you realize it's wasted energy. So I stopped. I remained professional, but I let go of the idea that this had to be more than what it was. I had been working toward an outcome the other person wasn't, and that's okay.

I've also seen organizations persist too long with underperforming employees, moving too slowly to begin the "manage up or manage out" process. The intention is often good, give someone another chance, another training, another cycle, but when years go by without change, the cost to the team can be high. In these cases, the perseverance is not grit, it is avoidance. The wiser choice is to step in decisively and create the conditions for either growth or transition.

It's easy to wear struggle like a badge of honor. But sometimes, perseverance can blind us. It can trap us in a cycle of effort that no longer serves us. The entrepreneur who refuses to pivot a failing idea. The athlete who plays through an injury until it forces them out for the season. The person clinging to a job, a relationship, or a belief that no longer fits who they are.

Without reflection, perseverance becomes *grit without growth.* The real challenge isn't just pushing forward: it's knowing when to persist and when to pivot. It's having the courage to ask: Am I enduring something that strengthens me, or am I resisting a change I need to make?

Leaders should build in periodic reviews to ask: Does this goal still align with who I am and what we are building? Perseverance is noble, but only when it is anchored to purpose.

How to Become More

Perseverance is persistence with wisdom. Use these prompts to examine whether you are pressing forward or resisting a needed change.

- What has tested your perseverance the most?
- Has there ever been a time when perseverance became stubbornness?
- Where in your life can you reframe a challenge, not just as something to push through, but as something to reassess?

Perseverance isn't just about enduring hardship. It's about coming out the other side, not just stronger, but wiser. Where resilience steadies us in the shock of setbacks, perseverance sustains us through the long road that follows. It's not just enduring hardship, it's finishing what still matters once the storm has passed.

One MORE Thing

Perseverance is powerful, but only when paired with reflection. Keep going; but make sure you're going in the right direction ... toward a purpose that still matters.

20 | THE POWER OF GENEROUS LEADERSHIP

How Giving is Shaping the Future of Leadership

I can still remember the first time someone offered me an opportunity they saw me capable of before I saw it in myself. It was 2000. I was 27 years old. VerticalNet was less than a year past its IPO, growing rapidly, bringing in executives to scale the business. One of them tapped me to become the Director of Training for CX and Sales. Looking back, it was a quiet act of **leadership generosity**, someone choosing to believe in me before I fully believed in myself.

That moment didn't just change my title. It changed my confidence. It helped me recognize the talent I hadn't yet seen, waiting for permission to grow. It's funny how often the moments that change our lives don't announce themselves. No grand gestures. No ceremonies. Just a small act of generosity, someone giving a little time, a little trust, a little opportunity they could've kept for themselves. That's what leadership looks like at its best: not bigger titles, but bigger impact on others. When we talk about leadership, we often talk about vision, execution, results. But rarely do we talk about generosity. And yet, in my experience, **generosity is the real multiplier**. It's the quiet force that builds trust, deepens loyalty, and creates opportunities far bigger than any one person could grab alone.

Several times in my career I've elevated someone into management before they believed they were ready. Each time, I watched a spark of belief ignite in them. My choosing them was seen as generosity, and in return they gave more of themselves. Generosity isn't favoritism or lowering standards. It's about opening doors for those who are ready to walk through them, even if they don't believe it yet.

Redefining Generosity as a Leadership Strategy

Generosity is often misunderstood. It's seen as soft. Optional. Nice to have. In reality, it's the strongest strategic lever a leader can pull. When you give your time, when you give credit freely, when you open doors without expectation, you're not diminishing yourself. You're multiplying your impact. You're creating a culture where trust, ownership, and shared success thrive. You're fostering a work environment that **Becomes More** ... more resilient, more human.

Generosity isn't charity. It's strategy.

It's how leaders build cultures that outlast them and produce more ... more contribution, more productivity, more kindness, more good. Wisdom hoarded is potential stranded. The same is true of opportunity. When leaders give freely, they release potential that otherwise stays locked inside one person. Generosity turns hidden capacity into visible contribution. If you want to become a more generous leader, it starts with how you spend your time, how you share credit, and how you open doors.

Here's how generosity takes shape in practice.

Three Forms of Leadership Generosity

Time.

In a world where everyone feels rushed, giving someone your full attention is an act of leadership. Sometimes it's ten extra minutes of mentoring. Sometimes it's staying in an All-Hands meeting after the scheduled end time, answering every last question. Sometimes it's listening, really listening, without rushing to solve. It's being fully present.

Credit.

A generous leader lifts others into the spotlight. They hand out credit like seeds, not rewards. Because when people are recognized, they grow. And when they grow, the whole organization grows with them.
Four to one ... four acts of praise for every one piece of criticism.
That's the ratio generous leadership strives for.

Opportunity.

Opening doors is where real leadership lives. Not offering a chance only when it's safe, but offering it when it matters, giving someone a shot you could have taken for yourself and cheering them on as they rise. And it's not only giving opportunities to the obvious candidates. It's about intentionally lifting up those who have been overlooked too often, especially women, whose talent, insight, and leadership have long been underrepresented in corporate America.

When you give someone else the mic, you're not giving up your power. You're proving you never needed to cling to it in the first place. Generosity is deeply rooted in Pillar O: Opportunity. Because every act of giving creates more room for others to rise. Leaders who give credit, time, and access create a culture where opportunity is multiplied instead of hoarded.

The Ripple Effect of Generosity

Every act of generosity creates a ripple. That ripple becomes trust. Trust becomes momentum. Momentum becomes a force that carries forward, long after you're gone.

The greatest leaders aren't remembered for their accolades. They're remembered, for who they helped.

For me, decades later, I still remember that first act of leadership generosity. I'd argue I may not be writing this right now without that giving. The leaders who give generously don't just create success. They create successors.

How to Become More

Generosity multiplies leadership. It builds trust, loyalty, and momentum. Use these prompts to reflect on where you can give more freely.

- Where can you give credit today that will lift someone else's confidence?
- Who around you needs an opportunity and how can you open a door?
- Are you leading for recognition or for the ripple effect?

One MORE Thing

To **Become More**, give more. True leadership is measured not in recognition but in the impact created when you choose to give freely and consistently. The future isn't built by those who keep score. It is built by those who give quietly and consistently, trusting that what they plant today will bloom in ways they may never see.

21 | THE WILL TO ENDURE: WHEN MOTIVATION BECOMES WHO YOU ARE

How Fear and Duty Become Motivation

Most people talk about motivation like it's a spark, something you ignite with the right book, the right speech, the right morning routine. The idea is that if you can just get excited enough, you'll find the energy to push forward. That's never been my experience. For me, motivation isn't excitement. It's endurance. It's never stopping, because stopping isn't an option.

When Motivation Becomes Duty

For more than two decades, my reason for doing anything has been simple:
- Provide financial security and prosperity for my family.
- Lower the barriers to success for my children.
- Model and teach them integrity, honesty, a strong work ethic, generosity, and kindness.

It's never been about passion. It's never been about chasing personal glory. My motivation is rooted in something much deeper: duty, responsibility, and a refusal to let the past repeat itself.

Because I know what real, personal struggle feels like. And everything I do is to prevent my kids from experiencing that. Don't misinterpret that statement though. I am not shielding my children from failure, poor decisions, loss, or challenge; as adversity is necessary to strengthen their character and build their resolve. However, growing up "without" makes personal achievement more difficult. It is this privilege, one of solid financial footing, that I aim to provide my children as the springboard upon which they may lead their lives.

I've been laid off twice in my career. Both times it felt like a gut punch, a ruthless reminder of how unforgiving business can be. Fear and doubt crept in, but the thought of my family quickly reset my focus. That's when duty takes over. Your will to endure challenge, hardship, and uncertainty becomes the energy of your motivation.

Lessons from Absence: When Motivation Comes From What You Don't Want to Be

When my father left, he chose himself over his children. I'll never fully understand his circumstances, and I'll never know how difficult that choice was for him. But I know one thing: his life won out over ours. I learned from that. And my response, maybe the defining motivation of my life, has been to do the opposite. To always choose others over self.

That's where my motivation comes from: not self-actualization, but the determination that no one I love will ever feel that kind of uncertainty.

Fear as a Driving Force

Most people run from fear. I've learned to use it.
- Fear of going backward keeps me pushing forward.
- Fear of complacency keeps me evolving, no matter how much I've achieved.
- Fear of becoming what I saw in my father forces me to show up, every single day.

I don't wake up feeling "motivated." I wake up knowing there's work to do toward that objective. And that's the difference between motivation as a feeling and motivation as endurance.

The Motivation No One Talks About

Most motivational advice focuses on passion, inspiration, and ambition. Here's a reframing:

- Not all motivation is positive. Sometimes, the strongest motivators are things you never want to experience again.
- Motivation isn't always loud. It doesn't have to be a fire. Sometimes it's just a quiet, relentless force that refuses to allow one to quit.
- Motivation isn't a choice. It's a duty. I don't *find motivation*. It's already decided. The mission is set.

I don't chase motivation.

I wake up and do, because it's become who I am.

This is where motivation, perseverance, and hope diverge. Motivation is the internal engine of the **MORE Leadership Framework**. Perseverance is the steady action of pushing forward. Hope is the belief that tomorrow can be better. You need all three, but motivation is what gets you out of bed before either perseverance or hope can go to work.

What Keeps You Moving When You Have Nothing Left to Prove?

For years, my drive was about proving I could build something, achieve financial stability, and create opportunities for my family.
Now, the challenge is different. I don't need to prove anything anymore. I've already built the life I once dreamed about.

But the fear still lingers.

- Not fear of failure, but fear of slipping.
- Not fear of what I haven't done, but fear of what could be undone.

And that's why **I keep going.**

Because when motivation isn't about personal gain, when it isn't about passion or status or excitement, when it's built on something deeper, **it never runs out.** When motivation becomes part of your identity, you no longer need to chase inspiration. You carry it with you. That's what makes it enduring.

How to Become More

Motivation that lasts is built on endurance and purpose, not excitement alone. Use these prompts to clarify what keeps you moving.

- Am I waiting to feel motivated, am I missing a deeper purpose?
- Is my motivation tied to external rewards, or is it built on something unshakable?
- Have I been avoiding fear, when maybe I should be using it as a source of momentum?

One MORE Thing

Motivation that depends on excitement will always fade. Motivation that's built on endurance never does. Because at some point, motivation isn't a feeling anymore. It's just who you are. Endurance turns motivation into identity. That is the will to endure.

22 | CURIOSITY'S ROLE IN DEFINING GREATNESS

Why the Willingness to Ask, Explore, and Challenge is What Separates the Best From the Rest.

I've always been curious.

It just comes naturally to me. It always has.

Growing up, I was the [annoying] kid in class whose hand was always raised, ready with another question. I wasn't trying to be disruptive; I just seemed to always have another idea that needed more feeding. I was *seeking to understand. To know why.* That instinct never left me. It shows up in my photography, my writing, all my creative work, and my critical thinking. It's the force behind the constant stream of ideas that I come up with: How was this built? How can this be redesigned? What can make this better? What is this metric truly telling us?

And now, I find myself fascinated by curiosity as a trait and skill. Intellectual curiosity is one of the more overlooked characteristics in hiring, yet I find it's one of the core differentiators that separates average performers from great ones. The insatiable desire to discover answers, to **keep asking why, to** push the boundaries of how things are done, and to find a better way to make it more. Constantly iterating, embracing failures as springboards forward, seeking lessons in our bad choices, examining our lives.

That is curiosity.

Think about the best people you've ever worked with, the ones who truly stand out. They aren't just good at executing; they question things. They want to understand how things work, and more importantly, how they could work better. They don't accept the way things have always been done; they lead us toward what is next. They look for what others don't see.

Intellectual curiosity fuels the relentless pursuit of improvement.

That said, in many organizations, curiosity is discouraged. Some leaders think they are supposed to have all the answers. Some cultures punish people who question the boss. The danger is that curiosity gets quietly shut down, and with it the chance for innovation. The challenge for modern leaders is to build environments where questions are welcome, even the simple ones. If leaders never admit they don't know, or never ask questions themselves, they signal to their teams that curiosity is unsafe. The best leaders model it by asking questions out loud and rewarding those who do the same.

The Limits We Set

Yet, as I reflect on this idea, I'm left wondering, "What are the dangers of curiosity?" Earlier, we acknowledged perseverance can sometimes blind us. And now we must consider what is the lesson to be learned from the fact that: *"Curiosity killed the cat."*

Cats are known for their agility, resourcefulness, and independence. And yet, curiosity kills the cat. Maybe it's meant to suggest that we need a barrier to our curiosity, a limit to how far we should go. Perhaps it's a cautionary tale about risk, a subtle warning about venturing too far into the unknown, where uncertainty reigns and guarantees vanish. Perhaps it's society's way of telling us to stay within certain bounds, not to dig too deeply into things that might unsettle others or shake the foundations of the status quo. But placing artificial limits on curiosity comes at a cost.

Because, I believe: All growth, personally and professionally, lies just beyond the comfortable edges of what we already know. Every meaningful innovation, every significant leap forward, has come from someone willing to push past these artificial boundaries. If we restrain our curiosity, do we also limit our potential? Curiosity may carry risk, yes, but the greater risk lies in not pursuing curiosity at all. It's in accepting limits imposed by others, in quietly accepting the idea that it's safer not to know, not to challenge, not to explore.

Leaning into the Unknown

Explore that boundary. Rather than shrinking from uncertainty, lean into it. Choose to question assumptions instead of accepting them at face value. Ask the difficult questions: the kind that may make you (and others) uncomfortable. And don't underestimate the power of simple questions and returning to the basics. Some of the greatest breakthroughs start with "Why do we do it this way?" or "What if we tried something different?" No question of a curious mind is too small, and leaders who ask the basics often get to the root causes others miss. Curiosity doesn't always lead to breakthroughs overnight. Sometimes it's about the 1% improvements — the small questions that compound into major change over time.

Because while curiosity might carry the danger of the unknown, it also holds within it the promise of discovering what others overlook. It's the skill of curiosity that will set you apart, help you lead more effectively, and unlock what might otherwise remain hidden: opportunity.

In the AI era, this trait becomes even more critical. Machines can deliver answers faster than ever, but humans must still pose the right questions. Curiosity is what keeps leaders relevant, innovative, and human. Without it, we risk becoming answer-takers instead of pathfinders.

How to Become More

Curiosity pushes beyond limits and fuels discovery. Use these prompts to explore how you bring curiosity into your daily work and life.

- Where in your daily work are you already applying curiosity without noticing it?
- What question have you avoided asking that could open new understanding?
- Where might curiosity help you see opportunity beyond what is comfortable?

One MORE Thing

The lesson in curiosity isn't that we should fear where it leads, it's that we should never stop following it. Machines may give us answers, but only curious humans can ask the questions that shape the future.

23 | VISION AS A SUPERPOWER

How to Focus, Set Direction, and Lead the Way

> You've got to visualize where you're headed and be very clear about it. Take a Polaroid picture of where you're going to be in a few years.
>
> -Sara Blakely - founder Spanx

> All great things begin with a vision... a dream... I've always believed that success comes from not letting your eyes stray from that target. Anyone who wants to achieve a dream must stay focused, strong, and steady.
>
> -Estee Lauder - founder Estee Lauder

These words from two extraordinary women capture the essence of vision: the power to see where you're going and to stay resolutely focused in the face of challenges.

Remember, vision isn't only setting goals. It is activating your superpower of focus, clarity, and the ability to see opportunities others don't. In a world moving at the speed of AI, the ability to articulate and pursue a vision is more important than ever. Technology can optimize processes, automate tasks, and provide answers at scale. But it cannot create vision. Only human leaders can define purpose and direction in the midst of chaos. That is why vision is the human superpower in an AI-first world.

"As you think, so shall you become." - Bruce Lee

This timeless idea reminds us that vision is more than what we hope for. It is how we see ourselves today. The way you envision your future influences your actions, your mindset, and ultimately, your reality. Vision is both deeply personal and inherently collaborative. While it starts with how you see yourself and what you believe is possible, its true power lies in how it extends outward to the teams, organizations, and communities you influence. Your unique perspective can shape not only your own reality but also inspire new opportunities and possibilities for those around you.

Vision cannot be a secret. A vision kept inside one person's head dies there. A true leader communicates vision openly and repeatedly, rallying people around it until it becomes a shared purpose. An important note to consider in relation to the vision of your organization, and that is: You don't need to be a CEO to contribute to your organization's vision. Visionaries exist at every level, people with curiosity, insight, and the courage to challenge assumptions. By sharing your ideas and perspectives, you can help shape the future of your team, organization, or industry, uncovering opportunities others might miss.

The pace of change is accelerating. Having a clear vision of who you are, what you want, and how you'll approach challenges will be critical. Vision is so much more than looking ahead; it's grounding yourself today so you can keep pace with tomorrow. In times of disruption, vision steadies people. It provides purpose when everything else is shifting too quickly to grasp.

Ending Pillar O with Vision is intentional. Opportunity without vision can scatter energy. Vision provides the focus that ensures curiosity, optimism, hope, and perseverance all move in the same direction. It is the trait that connects personal drive to collective progress, and the one that keeps leaders relevant in a future defined by change.

How to Become More

Vision is both clarity and focus. It allows us to see possibilities and move toward them with conviction. Use these prompts to clarify your own.

- **How do you define your vision for your future?**
- Is it rooted in personal growth, professional milestones, or a balance of both?
- **What steps will you take to align your daily actions with that vision?**

Consider how small, intentional changes can have transformative effects over time.

- **How can you prepare to thrive in a faster, more AI-driven world?**
- What tools, skills, or mindsets will help you keep up and stay ahead?
- **How can you contribute to your team or organization's vision?**
- What unique perspective can you offer that helps others see opportunities they may have missed?

One MORE Thing

Vision matters because it inspires others to see what's possible. In an AI-first world, vision is the uniquely human trait that set direction. Hold it steady, share it boldly, and invite others to help bring it to life.

BLANK – FOR NOTES

O | OPPORTUNITY

Remember, Red, hope is a good thing, maybe the best of things, and no good thing ever dies.

— Andy Dufresne, The Shawshank Redemption (1994 film)

Opportunity's role in the framework is to provide the unshakeable strength to endure the good times, the bad times, and the dark ones. Struggle is inevitable, struggle is life. The direction we take through it is in our control. Opportunity reminds us to head toward more.

O | OPPORTUNITY:
Optimism, Hope, Perseverance, Generosity, Motivation, Curiosity, Vision.

BLANK – FOR NOTES

R | RELATIONSHIPS

At the core of every enduring leader is one truth they understand: people do not follow strategy, they follow connection. Connection to a cause, connection to shared values, and connection to each other. It born of trust, accountability, and belonging. Relationships are what scale culture. They are what allow teams to endure setbacks, adapt to change, and thrive together.

In an AI-first economy, it will be tempting to reduce people to data points or transactions. But the leaders who last will be the ones who remember that relationships are not a distraction from the work. They are the work. A leader's purpose is to help those around them succeed before anything else. Connection is not about knowing everyone's story. It is about caring enough to listen, to notice, and to respond. It is about building environments where people feel seen and valued, and in turn, give their best.

In this pillar, we will explore six traits that deepen connection: Empathy, Social Awareness, Listening, Conflict Resolution, Sincerity, and Kindness. No leader succeeds alone.

R | RELATIONSHIPS
Empathy, Social Awareness, Listening, Conflict Resolution, Sincerity, Kindness

BLANK – FOR NOTES

24 | EMBRACE EMPATHY FOR GROWTH, INCLUSION, AND CONNECTION

Explore How Empathy Can Help You Lead with Inclusivity and Make a Greater Impact in Your Work and Life

Empathy: The Bridge Trait

Empathy is more than just understanding someone's feelings. It's the ability to pause, step outside yourself, and see the world through another's eyes. In leadership, it's the difference between managing tasks and truly connecting with people. Without empathy, even the best strategies fall flat, because people don't just follow plans, they respond to how they're seen, heard, and valued.

I've found that empathy shows up in moments both big and small: listening to a colleague under stress instead of rushing to solutions, recognizing an unspoken hesitation in a meeting, or making space for someone's story when time feels scarce. These aren't grand gestures; they're what happens when you place others before self. They're actions that build trust, loyalty, and belonging.

Displaying empathy isn't soft. It is kind. And it is strategic. It's what helps us build inclusive cultures, design better solutions, understand differences, and create environments where people feel safe enough to contribute their best.

Some leaders confuse empathy with agreement or weakness. Empathy doesn't mean lowering standards or excusing poor performance. It means understanding someone's perspective so you can lead them more personally, sometimes with support, sometimes with accountability.

Why Empathy Matters:

- **Empathy and Inclusivity**: To foster inclusivity, we need to recognize our own biases and understand the experiences of others. This makes empathy essential for creating workplaces and communities where everyone feels valued, no matter their gender, race, or background.
- **Empathy in Patient Care**: In digital health, empathy becomes even more critical. As we automate and innovate, patient-centric care must remain the priority. Empathy allows healthcare providers and innovators to understand the unique needs and challenges of patients, driving better outcomes and greater trust in new technologies.
- **Empathy and Innovation**: Empathy also drives innovation. By understanding the challenges of others, leaders can create solutions that make a real difference. Teams with empathy are more creative, collaborative, and better equipped to tackle complex problems.

In times of change, empathy becomes even more important. Consider how employees feel when new AI systems or processes are introduced. Pausing to acknowledge their fears, inviting questions, and listening to dissenting voices doesn't slow progress, it speeds adoption and strengthens trust. And remember, empathy is not only moral, it's strategic.

Teams that feel understood are more resilient, make smarter decisions, create stronger connections, and establish strong relationships. Human connection fuels success because it helps people adapt and thrive together.

How to Become More

Empathy is the bridge between understanding and action. Use these prompts to reflect on how you connect with others.

- What biases do I have that prevent me from showing empathy?
- How can I practice empathy more intentionally in my daily interactions?
- When am I most empathetic? Least? Why? What does that reveal?

Commit to one action that demonstrates empathy. Whether it's listening more attentively in a conversation, considering someone else's perspective at work, or even practicing self-compassion, focus on empathy in action. Observe how this impacts your relationships and your ability to collaborate.

One MORE Thing

Empathy is about building spaces where people feel understood and valued. It's strategic; and table stakes to lead today. Empathy builds trust, fuels innovation, and fosters belonging. Show up with empathy daily, and you'll lead with impact.

TRAIT: SOCIAL-AWARENESS

25 | WHAT THEY DON'T TEACH YOU ABOUT LEADERSHIP

Social Awareness is One of the Most Underestimated Leadership Skills of Our Time. Quiet in Nature, but Critical to Building Trust, Connection, & Performance.

The Space Between Us

Social awareness isn't just being "in tune" with others; it's understanding the dynamics at play in a room, a meeting, a message thread, or a moment. It's the space between people, what lives in tone, silence, posture, and energy. It's a shift from looking inward to looking outward, from what we think and create to how and what we see.

And today, it's harder to develop than ever.

There was a time when we left the house each day, and with it, left behind our mess, our stress, and our personal worlds. That separation gave work a structure. Now, for many, the office is the home too, and with it, the lines have blurred. The realities of life, kids, illness, grief, anxiety, caregiving, financial strain, aren't left at the door. There **is** no door to walk out and leave it all behind. And leaders who aren't paying attention to this shift risk missing what people actually need to perform at their best. Dashboards and metrics can tell you what is happening, but they can't always tell you why. Social awareness fills that gap by keeping leaders tuned to the human signals data can't capture.

Some assume social awareness is innate; that you either have emotional radar or you don't. But it can be learned and strengthened with practice. Leaders can develop this skill by deliberately scanning the culture: who speaks up, who stays quiet, and whose voices go unheard. By asking a trusted colleague what signals they might be missing. By paying attention not only to what is said, but to what isn't.

Everyone Brings a Story

Diversity isn't only visible — such as gender, race, or age. It's also invisible: the why behind each person's work.

For one person, work might be a path to validation. For another, it's just a paycheck, a means to something else. For someone else, it might be a lifeline. Or an escape. Social awareness invites us to pause and consider:

What might be true for this person that I can't see?

As Peter Drucker reminds us,

The most important thing in communication is hearing what isn't said.

That insight is even more relevant today, when so much of our interaction happens over screens and in message threads, where tone, tension, or hesitation can easily be missed or misread. Understanding what's unspoken is often the fastest way to unlock performance. Misreading it is one of the easiest ways to erode trust. Remote and hybrid work require new forms of social awareness. Without hallway conversations or casual cues, leaders must learn to read tone in text, watch for long silences in group chats, and create intentional spaces for informal check-ins that used to happen naturally in the office.

Reading the Room, Even Over Video

Emotions are contagious. If you're a leader, you're broadcasting all the time. Your tone, your posture, even the speed of your speech sends a signal.

But in a hybrid world, signals are fuzzier. Someone might leave a camera off because their toddler is sick. Someone else might seem quiet because they're overwhelmed, not disengaged. And a smile might mean "I'm okay," or it might mean "I'm holding it together the best I can."

Social awareness is the leadership skill that notices what others might miss.

It's grounded in empathy, compassion, and kindness. And it asks us to see our teams as more than roles or titles. It invites us to strip away the labels, to set aside assumptions, and to remember that every person is carrying something beneath the surface. Because before they are employees, they are human. And great leadership begins there. It's what helps us create space instead of pressure. Understanding instead of assumptions.

One practical exercise: in your next meeting, focus on the group instead of the content. Who speaks freely? Who hesitates? Who is cut off? This simple culture scan can reveal power dynamics, cliques, or biases shaping the conversation. Social awareness grows when you slow down enough to notice the whole, the community, not just the individuals.

Bandwidth and Belonging

This trait also helps you design better team environments. Some people are drowning in invisible responsibilities. Some are introverts who process before they speak. Some are navigating cultural norms that shape how they show up. Being socially aware doesn't mean lowering the bar. **It means meeting people where they are so they can rise to what's expected.**

Because when people feel psychologically safe, seen, heard, and respected, they contribute more. They grow faster. And they stay longer. That kind of safety isn't framed as being nice. It is building and creating the trust people need to bring their full selves to the table. As Adam Grant puts it:

Psychological safety isn't about being nice. It's about giving people the courage to speak up—because they know they won't be punished for saying something smart, or something stupid, or something that needs to be said.

How to Become More

Social awareness is noticing what others miss. Use these prompts to reflect on how you read the room and respond to others.

- Who on your team is most different from you? Do you truly know their story?
- Has your emotional state ever changed the mood of a meeting, for better or worse?
- When was the last time you asked a trusted colleague what you might be missing in group dynamics? Social awareness grows when you invite feedback.

One MORE Thing

Social awareness isn't fixed. It is learned. By caring enough to notice and practicing deliberately, anyone can sharpen this skill. In today's blurred, complex, hyper-connected workplace, noticing what others miss unlocks trust, fuels performance, and deepens human connection.

26 | THE LOST ART OF LISTENING

The Pace of Life Has Accelerated, but Our Need to Be Heard, and To Listen, Hasn't Changed.

The Lost Art of Listening

There was a time when listening wasn't rare. It was how wisdom was passed down from one generation to the next. Humans have traded that for speed.

Listening isn't passive. It's one of the most active, demanding, and underappreciated skills a leader must build. It's a discipline to be trained through habits.

Start with this simple experiment in your next meeting: talk last. Or spend an entire meeting asking questions without offering your response in return. Practice pregnant pauses. You'll immediately notice how much more others share once they see the space to share and begin feeling heard.

In a world that moves fast, rewards noise, and floods us with constant input, listening has become a lost art. But it's still one of the most transformative tools we have.

When done well, listening reshapes relationships, unlocks insight, and builds trust.

It's not only hearing. It is honoring. To listen well is to put your ego aside. To get curious. To give someone the dignity of your full presence. And that's exactly why it's so rare. Quite often, early in my career, I cut people off, convinced I already knew where they were going. When I finally trained myself to let them finish, I was surprised by how often I was wrong and by how much more my team shared once they realized they truly had the floor. We're conditioned to prove ourselves. To protect our point of view. To fill silences and fix problems. But listening isn't fixing, it's *witnessing*. It's creating space for someone else's truth, even when it's uncomfortable. Especially when it is.

Listening Beyond Words

Listening, at its highest level, goes beyond sound. It's intuitive. A pause. A shift in tone. A look away. A deep sigh. These are not just moments, they're messages. And the most emotionally intelligent leaders learn to read between the lines. To read body language.

Intuitive listening is about sensing what's not being said. Spotting burnout before it's spoken. Noticing when someone's energy dips, even when their calendar stays full. That kind of listening is powerful. And it's teachable, but only if we value it enough to practice. And that starts by asking one simple question:

What Are You Listening to?

Because in an era where everyone and everything is chasing your attention: where platforms are multiplying, algorithms are listening, and doomscrolling has become a default, what you listen to *matters.* If we're not intentional about what we're tuning into, we might not like what we're becoming.

Listening isn't just about others. It's what you allow in. The inputs you consume shape your focus, your perspective, and even your personality.

Why Listening Is So Hard

It's not just you. Listening is hard, for all of us. And the reasons fall into three core categories.

- **Mental interference** is the internal static: stress, anxiety, assumptions, and the impulse to jump in with a solution instead of sitting with someone's words.

- **Environmental overload** is the outside noise: pings, feeds, interruptions, and the constant flood of information. We're conditioned to multitask, which makes true focus feel like a luxury.

- **Human limits** are the physiological realities: our brains process faster than people talk, which creates space for distraction. Add in fatigue or burnout, and it's easy to lose the thread. And when leaders only half-listen, they unintentionally teach their teams that attention is optional.

Listening is also where other traits come together — mindfulness (being present), empathy (understanding another), and even emotional intelligence (putting aside your own need to speak). Practiced together, listening becomes the skill that integrates the rest.

Four Tips to Reclaim the Art of Listening

1. **Quiet the mind:** Before your next conversation, take a deep breath. Ask: *What am I bringing in with me and can I let it go, just for now?*
2. **Protect your attention:** Close the laptop. Flip the phone. Silence notifications. Signal presence. It matters more than you think.
3. **Pace with intention:** When your mind wanders, gently come back to their words. Let their rhythm anchor yours.
4. **Reflect on your relationships**: When's the last time you felt truly listened to? And when's the last time you offered that gift to someone else?

The pace of work is accelerating at a rate we've never experienced before. That is an undeniable truth. Mastering the art of listening - the act of listening - is your true currency today. *"How you spend your time is all you have?"* In that context, when you ask yourself that simple question, "**What are *you* listening to**?", my hope is that it has new meaning for you.

How to Become More

Listening is leadership's most underused skill. Use these prompts to reclaim the art of attention.

- When was the last time you felt deeply listened to?
- Who in your life needs you to listen more fully?
- What inputs are you listening to most often, and how are they shaping you?

One MORE Thing

Listening is a gift. In a world where everyone and everything is competing for attention, choosing to give it freely is an act of respect and care. Putting others before yourself by truly listening strengthens trust and deepens relationships.

TRAIT: CONFLICT RESOLUTION

27 | YOU CAN'T LEAD IF YOU CAN'T RESOLVE CONFLICT

Why the Hardest Conversations Are the Ones That Define Your Leadership

It's not who is right. It's what is right.

That one line, built into the leadership framework of my mentors, Rick and Terry Peterson, has helped me navigate almost every meaningful conflict I've encountered. It re-centers the moment. It shifts the goal from winning to understanding. From proving to improving. Because the fact is conflict is part of being human. We fight. We disagree. We see things differently. We have *feelings*, and yes, even at work, those feelings get hurt. Joining a meeting doesn't stop us from being human.

And yet, most of us were never taught how to navigate that reality. We were taught to smooth things over or push through. To keep the peace, even when something feels off. But connection, *real* connection, requires more than that.

It requires learning how to face friction, not fear it. To listen, not just respond. To resolve, not avoid.

Other traits in the framework like curiosity, creativity, and social awareness prepare us for conflict. So do empathy and listening. The same skills that help us understand others day to day become even more important when emotions run high. Listening fully and empathizing with the other person's perspective doesn't mean you agree with them. It means you value them enough to seek first to understand before you respond.

Most of us weren't taught how to resolve conflict. We were taught to navigate around it. Or to dominate it. But real leadership, at work or in life, comes from learning to engage with it productively. Avoidance may feel easier in the moment, but it always costs more later. Issues that go unaddressed grow into resentment, trust erodes, and teams start wasting time talking around the problem instead of with the person who could actually move things forward. I've seen leaders spend hours in side conversations, venting to everyone except the one person whom the conflict is with. That avoidance drains energy and quietly disrupts culture.

Here are three principles I've come to rely on:

1. Step away, then return

If you're emotionally triggered, don't force the conversation. One of the most important things you can do in conflict is to examine how your emotions are shaping your point of view. If you're seeing red, you can't see the truth.
Take space. Regulate. Return. Strength isn't always in staying. It's often in coming back better prepared.

2. "It's not who is right. It's what is right."

That mindset transforms conflict from a power struggle into a shared pursuit. It takes ego out of the equation and puts clarity in its place. So simple to say, often very difficult to do.

3. Ask: What part did I play in this?

This one takes courage. But every time I've paused to examine my role in a conflict, to hold up the mirror and look at what I said, how I said it, and what I assumed, I've found something worth owning. Ownership doesn't make you weaker. It's a sign of strength to admit when you're wrong, to acknowledge the part you played in creating the situation. Being honest with yourself is the first step toward understanding how your ego may have influenced the outcome.

In Today's Workplace...

Conflict has become quieter, subtler, harder to spot, but just as present.

It shows up in Slack threads, passive emails, skipped meetings, and quiet resentment. And in a remote or hybrid environment, it's easier to *avoid* than to *address.* But unresolved conflict compounds. It breaks trust. Kills momentum. Derails progress. Unresolved conflict consumes time and resources. Employees end up spending hours working around the issue instead of addressing it directly. Addressed directly, conflict can strengthen teams and even spark innovation, because conflict usually means people care about the outcome.

And not all conflict is the same.

Here are a few types I've seen most often:
- Interpersonal conflicts – Misunderstandings fueled by tone, timing, or misread intent.
- Hierarchy conflicts – When we treat those above us differently than those beside or below us.
- Employee-supervisor conflicts – Often rooted in mismatched expectations or unclear communication.
- Accountability and ownership conflicts – When it's unclear who owns the outcome, or when everyone thinks someone else does, or no one admits who made the mistake.
- Strategy / How-to conflicts – When we agree on *what* we want to achieve, but clash on *how* to get there.
- Legal or policy conflicts – These require structure, documentation, and sometimes external resolution.

- Customer complaint conflicts – When something went wrong, there's an opportunity to listen, empathize, and repair.
- Bias-based conflicts – Often invisible but deeply felt. When unspoken assumptions or inequities go unaddressed.

The key is not pretending conflict doesn't exist. You can't get by avoiding it. Every type of conflict, from misread tone to structural bias, needs engagement. Even strategy disagreements, which can be healthy, become destructive if left to simmer in silence. Each one requires something different. Some call for boundaries. Others call for listening. But they all start with awareness and the willingness to engage.

Remember the three principles:

1. If you're emotionally charged, step away. Come back with a clear mind and calm heart.
2. Focus on facts, ask, *"What's true for each of us?"* instead of *"Who's right?"*
3. Ask, *"What role did I play in creating this situation?"* If you're accountable, step up and own it. Self-examination is true courage.

How to Become More

Conflict cannot be avoided, but it can be resolved with care and courage. Use these prompts to reflect on your approach.

- When conflict arises, do I focus on who is right or what is right?
- How do I manage my emotions before stepping into a difficult conversation?
- What role did I play in creating or sustaining the conflict? Do I avoid addressing issues directly, talking to everyone else instead of the person who could resolve it?

One MORE Thing

Conflict is part of connection, and leadership means not avoiding it, but stepping into it with courage, clarity, and care. It can be uncomfortable. But so is growth. Without one, we rarely reach the other. Choosing resolution over avoidance is what defines true leadership.

28 | LEADERSHIP DOESN'T START WITH STRATEGY. IT STARTS WITH SINCERITY.

In a World Full of AI Written Polished Words and Scripts, Sincerity is How Leadership Remains Human.

With improved-quality AI-generated images, voices, and videos posted at increasingly faster rates, the lines between *what's real* and *what's AI* are blurring by the second. We used to wonder if someone was being sincere. Now we wonder if they're even *real!*

In a world where faces can be faked and voices cloned; sincerity may be one of the last unmistakable traces of being human: the signal of honesty, openness, and truth. And that's exactly why it matters more than ever.

In leadership, sincerity begins the work of building trust, inspiring others, and showing compassion. It's the bridge between what we hear and how we respond. The steady thread of consistency that says: *You can trust that what I say is true. And that who I am doesn't shift based on who's in the room.* It shows up when you're the same person in a 1:1 with a direct report as you are chatting with the CEO or mentoring an intern. When people experience the same you, not a version of you, they trust you. That's when leadership becomes seen and felt.

I'll never forget a moment with one of my team members. She had been delayed in responding to a message. When she did, she immediately apologized. Her reason? Her child had been sick and at urgent care. Yet, she felt the need to apologize. I could sense her tension and hesitation, as if she needed permission to be both a leader at work and a mom at home. I told her plainly: *Never apologize for being a mom*. She was relieved. But she shouldn't have had to hold her breath in the first place. That moment reminded me how sincerity in leadership isn't just about telling the truth when it's hard or offering clear feedback. It's about showing up as real people, with real lives, and giving others the space to do the same.

That's what makes sincerity essential, and not just another trait of the MORE Leadership Framework. Authenticity and sincerity are often used interchangeably, but they're not the same. Authenticity is about being true to yourself. Sincerity is about being true with others. Authenticity can sometimes slip into self-indulgence: 'This is who I am, take it or leave it.' Sincerity is steadier: it is honesty expressed in a way that earns trust, because people experience the same truth from you no matter the setting. We're bombarded with content that *sounds* right but doesn't *feel* real. Perfectly worded messages. Performative empathy. Leadership language written by algorithms. And now "Workslop," AI-generated work that looks polished on the surface but lacks the substance to meaningfully advance the task[1], is something leaders must confront.

Most employees know the difference. They can tell when a leader's words are scripted for optics instead of spoken from conviction. That gap erodes trust faster than any strategy mistake. But sincerity? You can't fake that. Not in a conversation. Not in a moment that matters. It's what makes people lean in. It's what builds trust when things get hard. And it becomes most obvious *in* the hard moments, when a real leader shows up with truth, presence, and care. That's why sincerity has to show up in feedback too. Not the scripted kind. The honest kind. The kind that **helps someone grow,** helps them **Become More**.

Negative performance conversations are never easy. But when you offer feedback with sincerity, not judgment, you give someone the chance to move forward. You also model what it means to care enough to be clear. One way to do this: replace judgmental language ("You failed at this") with forward-focused sincerity ("I want you to succeed, and here's what needs to change"). And then there's the version of sincerity that often gets left out: The kind you show when **you're** the one struggling. Sincerity means having the confidence to say, *"I'm not okay."* It means asking for help, being transparent about mistakes, and trusting those around you to lift when you can't carry it all.

As Eddie Vedder sings in *Wreckage*:

Even every winner has a losing streak.

Sincerity in struggle isn't weakness, it's credibility. When leaders admit they're in a losing streak, they give their teams permission to be real too. That honesty is what keeps trust intact when things are toughest.

How to Become More

Sincerity builds trust and signals truth. Use these prompts to reflect on how consistently you show up.

- Where do you struggle to be the same person across different audiences? What's keeping you from showing up consistently?
- When was the last time you gave honest feedback from a place of care? How was it received?
- Is there anything you're carrying alone right now that your team would gladly help you carry if you let them in?

One MORE Thing

Sincerity is honesty. Sincerity is truth. Sincerity is human. In a world of algorithms and endless polish, sincerity cuts through the noise. Showing up as your true self inspires trust and invites the same from others.

TRAIT: KINDNESS

29 | THE IMPACT OF INTENTIONAL KINDNESS

Explore How Small, Intentional Acts of Kindness Can Strengthen Bonds and Build Meaningful Connections.

Kindness is a trait that lies within most of us, but how often do we express it intentionally? In the midst of busy days and complex responsibilities, it can become an afterthought. Yet, when practiced with intention, kindness transforms not only individual interactions but the culture around us. Some leaders worry that kindness signals softness or lack of toughness.

But kindness isn't the opposite of strength. It's how strength is delivered. You can be kind and still uphold high standards, make hard calls, and hold people accountable. In fact, avoiding discipline or feedback is less kind, because it leaves people in the dark. True kindness is being clear, honest, and respectful, even when the message is difficult.

Reflection: Empathy in Action

When we covered empathy, we explored its power and how it allows us to connect with others in a meaningful way. During a trip to Las Vegas, I had the opportunity to put empathy into action in a very personal way. My father, who was absent for most of my life, passed away that year. His widow, a woman I had known as my "Aunt Sue," (which is another story in and of itself) lives in Las Vegas. She's now facing the loss of her husband in near isolation, with few friends or community connections. Despite our complex history, I chose to visit her, to be present, show kindness, and share a hug.

Setting aside past decisions and any lingering resentment, I focused instead on her loss, hoping to offer comfort in her grief and loneliness. This experience reminded me how kindness and empathy can allow us to bridge past hurts and offer understanding, even in challenging circumstances. It was an emotional visit, bringing back memories more than 40 years old, yet independent of how difficult it may have been for me, it pales in comparison to the loss my Aunt Sue is feeling.

That's why this chapter comes after conflict resolution. Kindness doesn't mean avoiding conflict. It means engaging in it with empathy, clarity, and care. Being kind and firm go together. Delivering tough feedback with empathy is far kinder than staying silent and letting someone fail later. Let's build upon empathy's foundation by focusing on intentional acts of kindness. The small, purposeful steps we take to bring connection and positivity into the lives of others.

The Power of Small Acts

During research for this book, I conducted a "mini social experiment" by choosing to express kindness intentionally in all my interactions. At Starbucks, for example, I noticed a few people struggling with payments. Without hesitation, I stepped in with my tap to cover their coffee. I didn't do it for recognition or thanks but simply to extend kindness in an unexpected moment. In one instance, Starbucks paid it forward back to me and bought my coffee in return, creating a small ripple of kindness that turned a routine frustration of one customer into an experience something more memorable for all of us. These moments not only brightened my day but also fostered a sense of connection that stayed with me long after.

In our daily lives, it's easy to overlook the small gestures, a helping hand, a quick word of encouragement, or even a simple smile. But when we approach these interactions with intentional kindness, they take on new meaning, creating an environment where genuine connections can thrive.

Strengthening Bonds that Stand the Test of Time

I also reconnected with a fraternity brother during my trip; someone I hadn't seen in over twenty-plus years. Despite the decades apart, our friendship is the type that when we reconnected, it was as if no time had passed at all. Incredible really. On reflection, I realized that what built our bond all those years ago was the kindness he showed to me from the moment we met. His early expressions of care, interest, and support laid a foundation for a friendship that remains as meaningful today as it was then. This experience reminded me of a personal quote I've held close:

True love never dies.

Intentional kindness not only builds new connections but also reinforces relationships that endure over time. It's a reminder that the bonds we create through kindness have a lasting impact on both our personal and professional lives. The same is true in leadership. When people look back on the leaders who shaped them most, they rarely remember quarterly targets or polished strategies. They remember the moments of kindness that gave them confidence, the tough conversations that were handled with care, and the leaders who made them feel seen as people first.

Earlier we explored community as the anchor of leadership. Kindness is how we build it, act by act, choice by choice. And it's fitting that kindness closes the Relationships pillar. Because after empathy, conflict resolution, and sincerity, kindness is what ties it all together. It's the daily practice that reminds us leadership is ultimately about being better humans with others.

How to Become More

Kindness costs nothing yet it changes everything. Take a moment to reflect on kindness and bring more intention to expressing it.

- When someone expressed kindness toward me, how did it make me feel?
- What stops me from expressing more kindness often?
- How might kindness help me strengthen relationships at work? At Home? In your community?

Look for small opportunities to practice intentional kindness and notice how it shapes your interactions and connections.

One MORE Thing

Never underestimate the impact of a kind heart. Kindness is not weakness. It is strength expressed with care. When practiced with intention, kindness has the power to transform ordinary interactions into deeper connections. Small acts of kindness create ripples that last far beyond the moment.

Kindness costs nothing, yet it can be so impactful and so meaningful, especially when given without expectation or debt.

former colleague, friend forever

BLANK – FOR NOTES

R | RELATIONSHIPS

A life is not important except in the impact it has on other lives.

— Jackie Robinson

Relationships are the fabric of life. They bind communities and organizations together, sustain families, strengthen teams, and shape fleeting moments with strangers. The MORE Effect calls us to choose kindness over indifference, empathy over neglect, listening over silence. In those choices, we determine the impact our life will have on others.

R | RELATIONSHIPS

Empathy, Social Awareness, Listening, Conflict Resolution, Sincerity, Kindness

BLANK – FOR NOTES

E | Execution

Ideas inspire. Vision guides. But only execution creates impact. Execution is about turning intent into progress. It is not speed for its own sake. It is the discipline of meaningful motion: the ability to move forward, adjust in real time, and carry energy through uncertainty.

This pillar demands courage, critical thinking, and adaptability. It asks leaders to balance urgency with patience, creativity with clarity, ambition with temperance. In an always-on world, contribution and productivity is what separates those who remain stuck in planning from those who shape the future.

In this pillar, we will explore six traits that drive contribution: Courage, Critical Thinking, Decision-Making, Creativity, Adaptability, and Temperance.

This pillar reminds us that leadership is measured not only by what we dream but by what we deliver. To execute with consistency and care is to leave a mark that endures.

E | EXECUTION
Traits: Courage, Critical Thinking, Decision-Making, Creativity, Adaptability, Temperance

BLANK – FOR NOTES

30 | COURAGE - THE STRENGTH TO BE YOURSELF

Embracing Courage Means More Than Bold Actions. It Means Standing Firm in Your Values, Contributing Authentically, and Finding the Strength to Grow, Even When it's Uncomfortable.

Reflecting on Kindness

To end the last section, we explored kindness, a trait that serves as a bridge, fostering trust and connection. Kindness opens doors, offering understanding and warmth in a way that few other qualities can. But while kindness builds bridges, courage is what enables us to cross them, especially when the path ahead is uncertain. Now we turn to courage, particularly the kind that asks us to stand firm, even when it may challenge those connections.

Everyday courage shows up more often than we think: giving honest feedback, admitting a mistake, or advocating for a bold idea in the face of resistance. These aren't headline-making moments, but they define leadership more than dramatic acts of bravery ever could.

The Foundation of Courage in My Life

Courage has been a core value of mine since my early twenties, when my mother had reframed my childhood nickname into my compass: Courage, Hope, Optimism, and the Pursuit of All Three. It's guided me ever since. For me, courage has always been the anchor that holds me steady — the inner resolve that keeps me moving forward, especially in the face of obstacles. It's important to note how courage differs from resilience. Resilience is about enduring hardship over time. Courage is about initiating action in the face of fear. Resilience helps you keep going. Courage is what gets you started when fear tells you not to move.

The Quiet Courage of Self-Examination

When we think of courage, we often envision acts of bravery, like soldiers who face danger for a greater cause. This is the most visible form of courage, a willingness to face external challenges. But there's another kind of courage that's equally challenging: the courage to turn inward. It takes real courage to examine ourselves honestly, to recognize our flaws, and to confront the parts of ourselves that may hold us back.

This internal courage is essential for growth. It requires a willingness to change, evolve, and continually strive to **Become More**. True courage means not only standing strong in the face of external challenges but also having the humility and resilience to face the difficult truths within ourselves. The courage to reflect and admit mistakes, to examine our flaws, to recognize our own limitations, and to pursue growth is at the heart of self-development and **Becoming More**. Courage here overlaps with vulnerability: saying "I need help" or "I was wrong." Fear is natural in these moments. Courage doesn't mean the absence of fear; it means acting despite it.

Courage at Work

Courage is a critical ingredient in today's workplace. Whether it's having the courage to speak up and contribute an idea, or the courage to make tough decisions as a leader, each form of courage plays an essential role in thriving amongst your peers in the workplace. This is particularly true for the emerging voices of Gen Y and Gen Z, who bring fresh perspectives and are eager to make meaningful contributions. As AI reshapes work, courage will also mean speaking up for the human element. It will take courage to make bold changes, to champion people over process, and to resist the pull of what feels like the safe status quo. Leaders will need to step forward even when technology tempts them to step back.

Courage to Contribute

Every person, regardless of position, requires courage to lean in, contribute insights, share ideas, and offer perspectives toward the outcome of the moment. This is particularly relevant for the emerging voices of Gen Y and Gen Z, yet knowing how and when to do so can be challenging, and thus requires courage. But meaningful contribution requires a deeper question: Am I speaking for myself, or for the audience and for the outcome? True courage to contribute means aligning our voice with a purpose beyond personal validation.

When we focus on the greater good, framing our ideas in terms of the outcome rather than ourselves, the act of speaking up often requires less courage. By shifting our mindset to serve the mission, the purpose of the project or team, we relieve ourselves of the worry or self-consciousness that can hold us back. It's no longer about the personal risk, but about what benefits the whole, not ourselves.

Whether you're new to the team, an entry-level employee, or a mid-level manager, it takes courage to bring forward your perspective or propose a new approach, especially when senior executives are present. But by focusing on the audience and the outcome, we shift our contributions from personal gain to the greater good of the organization. When we have the courage to offer insight with intention, we create a culture where every voice is valued, and ideas are shared with purpose. It's not just about speaking up, but about making a positive impact on the mission. This courage to contribute thoughtfully is essential for fostering a workplace where growth and meaningful progress are possible for everyone.

Courage in Leadership: The Courage Not to Be Liked

For leaders, courage often means making decisions that may not win immediate popularity. This is the *courage not to be liked*. A strength that asks us to stay true to our values and vision, even when it may disrupt harmony or risk approval. Leadership requires making tough choices and holding firm to a long-term mission, prioritizing integrity and purpose over short-term satisfaction. This form of courage isn't always comfortable, but it's essential for authentic leadership.

That's why courage comes first in the Execution pillar. Courage gives all the other traits a foundation. It's what turns principles into action. Without courage, empathy stays an idea, integrity stays a value, and vision stays a dream. Courage is what moves them into the real world. At its heart, courage is the strength to be yourself: consistent, value-driven, and willing to stand firm no matter the moment.

How to Become More

Courage is the strength to stand firm in your values, even when it costs you. Authentic leadership begins there. Use these prompts to reflect on how you show courage.

- How can you cultivate your courage? Do you have an unpopular idea you believe in, and how can you advance it?
- In what areas of your life (or work) do you feel the need to stand firm, even if it means standing alone?
- Where might your insights add value, and how can you approach these moments with the courage to make a difference?

One MORE Thing

Courage is the strength to act despite fear, necessary to face the struggles you will inevitably encounter. Standing firm when the ground is shaking demands bravery and defines modern leadership.

31 | BEYOND WHO'S RIGHT TO WHAT'S RIGHT

How Questioning the Status Quo and Breaking Down Bias Drives Innovation and Resilience in Today's Workplace

Critical thinking has been a cornerstone of human progress for thousands of years. From Socrates' probing questions in ancient Athens to the structured logic of Enlightenment thinkers, the ability to question assumptions and analyze information objectively has always been essential for advancing society. In today's always-on economy, it has become an indispensable skill for navigating complexity and fostering innovation in the workplace.

But many leaders overestimate their ability to think critically. We all fall prey to cognitive traps: groupthink that silences dissent, confirmation bias that makes us only seek evidence that supports our beliefs, or the urge to jump to solutions before fully understanding the problem. Recognizing these patterns is the first step to improving.

Let's explore three reasons why critical thinking remains as vital as ever and how we can ensure it thrives in modern work environments.

From Ancient Philosophy to Modern Necessity

The roots of critical thinking stretch back to ancient Greece, where Socrates developed a method of questioning that challenged the beliefs of his time. His approach laid the groundwork for Plato and Aristotle, who formalized logic as a tool for uncovering truth. Fast forward to the Enlightenment, and thinkers like Descartes, Locke, and Kant further developed the principles of reasoning, skepticism, and evidence-based thinking. Their ideas helped fuel revolutions in science, politics, and human rights, proving that critical thinking isn't just an intellectual exercise. It's a transformative force. The world was in fact not flat.

In the modern workplace, this tradition continues. **Critical thinking enables us to evaluate decisions, solve complex problems, and adapt to rapid change.** It's not only knowledge; it's how we use that knowledge to question, improve, and innovate.

Critical Thinking: Beyond Who's Right to What's Right

Workplace decisions are often built on legacy ideas, past policies, or outdated norms. Critical thinking challenges us to evaluate these foundations. Are they still relevant? Do they hold up to scrutiny under today's circumstances?

This process shifts the focus from ego, who's right, to collaboration and improvement, what's right. Seeking truth over ego is what separates modern leaders. They cultivate environments where debate is about ideas, not personalities. The goal isn't to win the argument but to find the best path forward.

It empowers individuals and teams to:
- Question long-standing practices.
- Validate decisions through evidence and logic.
- Foster innovation by rethinking the "how" and "why" behind systems and processes.
- Ask, "Why not?"

Critical thinking asks: Is this policy, procedure, or system designed for the company's future or clinging to its past?

Overcoming Bias and Rigidity

Even with the best intentions, biases both conscious and unconscious, can shape our thinking. Similarly, rigid adherence to tradition often stems from fear of change or over-reliance on established norms.

Critical thinking provides tools to identify and dismantle these barriers:

- **Breaking Bias:** It encourages us to question our assumptions, seek diverse perspectives, and challenge stereotypes.
- **Defeating Rigidity:** By promoting curiosity and adaptability, critical thinking helps teams move beyond "we've always done it this way."

Courage and critical thinking go hand in hand here. It takes courage to challenge consensus or propose a different view when everyone seems aligned. But without that courage, teams risk drifting into groupthink, where harmony is prized over truth and innovation stalls. Take hiring practices as an example. A critical examination of job descriptions or interview processes might reveal unintentional biases that hinder diversity and inclusion. Addressing these issues can lead to fairer and more effective outcomes.

Navigating the Reality of Workplace Culture

While critical thinking is essential, not every workplace is prepared for it. Some cultures and supervisors may resist open discussion or challenges to the status quo. This reality offers two important lessons:

For Leaders: It's our responsibility to cultivate a safe environment where critical thinking can thrive. This means creating space for open dialogue, valuing improvement over ego, and encouraging innovation. This includes welcoming constructive disagreement. Leaders who shut down debate miss blind spots; leaders who invite it sharpen collective intelligence. Building this kind of culture takes time, trust, and fortitude. But it's worth the effort.

For Employees: In environments less receptive to critical thinking, approach challenges with tact. Thoughtful and respectful communication can open doors to dialogue without triggering defensiveness. Your approach matters. Remember the outcome: what is right.

And in an AI-first world filled with generative AI outputs , this trait becomes even more critical. Algorithms can generate answers at speed, but only human leaders can question the output, spot flaws or hallucinations, and ask whether the recommendation makes sense in context. Critical thinking is how we prevent blind reliance on machines and ensure technology serves people, not the other way around.

How to Become More

Critical thinking cuts through noise and reveals truth. Use these prompts to sharpen your practice.

- How can the historical principles of critical thinking inform your work today?
- As a leader, what steps can you take to create a culture that encourages questioning and innovation?
- How can you create an environment where others feel safe to think critically?

One MORE Thing

Critical thinking is one of the most powerful tools for growth and adaptation. It gives fuel to innovation by separating truth from assumption and ideas from ego. In an AI-driven world, it's the human trait that questions, verifies, and ensures progress when change is constant.

TRAIT: DECISION-MAKING

32 | CLARITY UNDER PRESSURE: THE DISCIPLINE OF DECISION-MAKING

How Great Leaders Build Cultures of Confident Decision Makers

There are plenty of frameworks out there: OODA loops, Eisenhower matrices, RACI charts. One book I reference frequently has 50 models. But most days, I fall back on something simpler: the Ben Franklin pros and cons list. Because for me, I find my best decision making comes down to a simple formula: **Intuition + facts = informed choices.**

Studies suggest we make more than 30,000 decisions each day. Most are minor, what to wear, when to respond, what to eat, but others carry real weight. And when you're making decisions at that volume, decision fatigue is real. The more decisions you're forced to make, the more the quality of those decisions can deteriorate over time. That's why simplicity, clarity, and trust in your team are essential, not just for effectiveness, but for endurance.

And discipline in decision-making also goes beyond the choice itself. It continues with the follow-through. A decision isn't truly made until it's communicated clearly, carried out consistently, and evaluated afterward.

Leaders must ensure everyone understands both "the what" and "the why" so the organization can execute effectively. At the pace we work today, especially in an AI-accelerated environment, what leaders need to get comfortable with is the fact that: **you're not going to get every decision right**. That's not the goal.

Generally, I've found those who believe it to be the goal struggle with analysis paralysis which is even more costly today. Tools can surface endless options or data points, but leaders must still make judgment calls. Decision discipline means knowing when to stop analyzing, weigh intuition alongside facts, and move forward without being paralyzed by too much information. The real question is: *what happens next?*

- How do you handle a poor or wrong decision?
- Do you blame others?
- Clam up?
- Hide?
- Or do you take ownership and accountability?
- Do you adapt?
- Do you bring your team in or shut them out?
- Do you admit your mistake?

Because what builds a high-performance culture isn't perfection. **It's resilience**. It's what you do *after* the decision that matters, how you treat setbacks and how you lead through them, that sets the tone for your team. That also means having the courage to reverse course quickly when a decision proves wrong. Admitting mistakes and adjusting fast is far better than clinging to a bad call out of pride. The discipline is in learning, not defending.

Decision Making Is a Leadership Responsibility

Decision making isn't just a leadership skill. It's a leadership responsibility. Your job is to empower others to decide, too. To help them understand how critical it is they have the confidence to choose ... encouraging them to ask for forgiveness, not permission. Too many organizations still operate on fear.

People stay frozen, paralyzed by the fear of making the wrong move. But that's a failure of leadership, not of the individual's ability.

In a modern, high-trust team, performance isn't judged by a single decision. It's shaped across a series of moments: through consistent ownership, courage, and growth. As a leader, your role is to create a culture where people are safe to choose and supported when they need to course-correct. That culture also requires leaders to manage their own emotions under pressure. Taking a breath, recalling your principles, or grounding in your team's values helps steady the moment so your decision is guided by clarity, not stress. Even a pause of a few seconds before speaking can reset the tone and keep you from reacting emotionally rather than responding intentionally. That's how momentum is built. That's how organizations scale. And achieving that culture starts here: hiring.

Hiring Is One of the Most Critical Decisions You'll Make

Of all the choices you make, few are more consequential than who you hire. Because you're not just selecting a resume, you're shaping the future of your team. You want people who align with your values, work ethic, and principles. People who can think clearly, move decisively, and thrive in an environment where failure is treated as feedback, not punishment. **A person's character is as important as their credentials or past performance.** Hire for integrity, humility, adaptability, and temperance. Doing so offers a sense of calm, especially when you're not in the room. Because you're not building a team to execute tasks. You are building one to make decisions. You need a culture of confident, empowered decision makers.

Steps to Build Decision-Making Trust

1. **Define the decision owner.** Clarity beats consensus. Make it clear who owns the call.
2. **Align on the problem.** Before solving anything, make sure everyone agrees on what's being solved.
3. **Invite challenge.** Strong teams disagree out loud. Create safety for hard questions. Ask, "Are there any dissenting opinions among the group?"

4. **Be honest about trade-offs.** Every choice has a cost. Name them, don't hide them.

5. **Close the loop.** People commit when they understand the why, even if it wasn't their idea. Clarity also means communicating the decision well. A decision isn't truly complete until people know exactly what was decided, why it matters, and what role they play in carrying it out.

6. **Evaluate outcomes.** Revisit the decision. Evaluate outcomes together and treat them as feedback. Every decision is a chance to refine judgment and improve the discipline for next time.

The more decisions you decentralize and demystify, the less decision fatigue you'll see. High-performing cultures don't just make good decisions. They design for better decisions at scale.

Some Decisions *Do* Deserve More Weight

Most decisions are recoverable. But some, like hiring, investments, taking risks, and how you treat your people, carry long-term impact.

These are worth slowing down for. Worth seeking input. Worth asking, *"Will this hold up in 6 months?"*

A strong decision-making culture knows the difference between decisions that require speed and those that demand thought and consensus.

How to Become More

Decision-making is clarity in action. Use these prompts to reflect on how you choose and how you empower others to choose.

- Are my decisions modeling the values I want others to see?
- Have I created a culture where people feel safe to decide without fear?
- Where can I delegate decision-making and trust the outcome?

One MORE Thing

Leadership is not making perfect decisions. It's building a culture where choices are made with clarity, courage, confidence, and care. Growth comes not from avoiding mistakes but from how quickly you adjust, learn, and refine your discipline for the next decision.

33 | MORE THAN IDEAS. THE TRUE WORK OF CREATIVITY.

How Creative Thinking Unlocks Better Decisions, Smarter AI, and Stronger Outcomes

Ironically, when I sat down to write about creativity, I hit a wall: writer's block. The MORE Effect is built on one principle idea: embracing **More** goes beyond improving what you already know and expands into growing in ways you never imagined possible.

It's a high bar — *ways never imagined possible.*

Attempting to define creativity in this context stifled the very concept itself. Earlier caution was introduced with curiosity, and the potential blinding nature of perseverance was highlighted. Creativity too requires a fresh perspective.

From Curiosity to Action

With **curiosity,** the emphasis was on the drive to ask, explore, and challenge. But curiosity on its own is just wandering. Creativity is curiosity's direction.

Curiosity asks, *What if?*
Creativity says, *Let's find out.*

Thinking Differently About Creativity

Most often, creativity gets boxed into art, music, or design. But at its core, isn't it just problem-solving? It's the way you connect ideas, structure a pitch, analyze a report, or find a way around a dead-end. It's not simply coming up with ideas. It is making them work. Think about the everyday business scenarios: improving a clunky onboarding process, reframing a customer service script so it feels more human, or reshaping how meetings are run so people leave energized instead of drained. These are acts of creativity, even if no one calls them that.

The best creative thinkers aren't just dreamers. They're builders. And this isn't about sudden genius. More often, creativity is the discipline of experimenting, adjusting, and persisting until something useful emerges. It's a habit of discovery.

Aligning Work with Your Whole Self

For years I caught myself saying, "I'm only using half of myself." It wasn't a knock on the company I'd helped build. It was an honest read of my own capacity and an awareness that more of me was waiting to be used. That realization became a turning point. Creativity isn't just problem-solving; it's alignment. It's finding the spaces, inside your current role or beyond, where your strengths, values, and ambitions actually show up in the work.

If you feel under-expressed, begin with a candid conversation. Name what you believe you can contribute. Map your strengths to real needs. When leaders create environments that are safe for fresh approaches and critical thinking, creativity stops being a side project and becomes how questions are asked, problems get solved, and work gets done.

Creative Thinking in Analysis

Creativity isn't just about generating ideas; it's also making sense of them. Some of the most creative work happens in analysis, where patterns emerge, meaning is extracted, and new insights take shape. The best analysts, whether in business, research, or everyday decision-making, aren't just crunching numbers or organizing data. They're asking, *What does this mean?* and *What's the best way to interpret this?*

It's a process of framing, connecting, and synthesizing: turning information into something useful, actionable, and often unexpected. Creativity in analysis means seeing beyond what's obvious and constructing narratives that make the complex understandable.

These days, we're constantly asked to do more with less. This challenging scenario requires a blend of analytical and creative thinking. One must find innovative solutions to efficiently and effectively achieve business outcomes without sacrificing quality. The best operators excel by combining these skills, proving that creativity is essential even in analytical tasks. It's this fusion of creativity and analysis that enables us to navigate complexity and drive success.

Bridging Analysis and AI Prompt Writing

At the intersection of creativity and analysis lies a new frontier: **AI prompt writing**. Just as a great analyst crafts narratives from data, a well-constructed prompt directs AI toward meaningful, nuanced, and actionable responses. AI doesn't think for you: it responds to how well you frame the question. A vague prompt gets a vague answer. A precise, well-structured prompt unlocks insight.

The creativity isn't in AI itself. It is in how you use it.

AI Prompt Writing: A Modern-Day Form of Creativity

As AI becomes further embedded in how we work, prompt writing is emerging as a critical skill. Asking questions isn't enough; one must develop the right kind of input to produce better, more relevant outcomes. The most effective people will be those who combine analytical thinking with creative prompt crafting, using AI as a tool to expand their thinking, refine their ideas, and move faster toward solutions. This is creativity in its modern form, not just imagining what's possible, but engineering better results through smart, intentional input.

Reframing Creativity for the Modern Workplace

Creativity isn't a luxury reserved for artists. It's not a side project or an "extra" skill. It's embedded in how we solve problems, analyze data, and even interact with AI. It's also what sets apart those who simply react from those who actively shape the future. Leaders who lean into creativity don't just generate ideas — they give their teams permission to test, learn, and adapt. This builds cultures where innovation comes not from one person's brilliance but from collective execution. And sometimes, creativity means pushing through the moments when you feel stuck.

This chapter was born through writer's block. The irony wasn't lost on me. Creativity rarely arrives in a flash. It reveals itself in motion, in the courage to take the next step before the perfect idea appears. And that's true whether you're solving a business problem, refining an AI prompt, or just trying to move an idea forward. So the real question is: How can we be more intentional about creativity in our work?

How to Become More

Creativity is curiosity with direction. Use these prompts to bring more creativity into your work and leadership.

- Where in your work do you apply creativity without realizing it?
- How can you blend analytical thinking with creative problem-solving to drive better outcomes?
- How can you reframe challenges as opportunities to exercise creativity in unexpected ways?

One MORE Thing

Creativity is more than what you can imagine. It's the action you take afterward. If you feel stuck, remember: *the way forward isn't waiting for the perfect idea, it's starting with what you have.*

34 | ADAPTABILITY – THRIVING AMID CHANGE

The Quiet Force of Survival and Growth

Adaptability is a foundational trait for excelling in the workplace today: the ability to manage fast-paced, diverse schedules, connect effectively with others, and stay calm and focused in uncertain environments, when things go wrong. The modern workplace demands more than expertise or hard work. It calls for the **ability to adapt** as table stakes. To meet every moment with clarity, and to move forward with purpose, *adaptability* as a mindset.

Adaptability also means acknowledging the emotions that come with change, fear, loss, or discomfort and learning how to move through them. Change rarely feels easy, but growth often begins in those moments of discomfort. Adaptable leaders model this by getting comfortable being uncomfortable, showing their teams that uncertainty is not something to fear, but something to navigate together.

Adaptability Through Context Switching

If your schedule looks like a string of back-to-back meetings, you've likely felt the challenge of switching between subjects, teams, and priorities. Adaptability here isn't solely survival; it extends to ***mastery***. Can you clear your mind between calls, recalibrate for the next topic, and show up fully present? This takes more than multitasking. It requires intentional resets, mental agility, and rapid context switching. Even a brief moment to pause, to breathe, and to ground yourself before diving in can help you bring clarity and focus to every conversation.

Adaptability in Communication

As leaders and performers, how often do we stop to consider *how* we communicate, not just *what* we're saying? Adaptability in the form of communication means meeting people where they are: adjusting your tone, language, and approach to ensure they understand, hear, and act.

It's recognizing that every person or group you speak with processes information differently and tailoring your delivery accordingly. Whether you're leading a team, presenting an idea, or offering feedback, success often depends on how well your message resonates. This kind of adaptability creates clarity, alignment, and trust. The most effective servant leaders keep this top of mind, as they know it is more important to speak for the audience than oneself.

Adaptability to Uncertainty and Change

Change is inevitable, but our response to it is what defines us. For leaders, adaptability isn't only about adjusting plans; it's about keeping others calm in the storm. Teams watch how leaders respond. If you panic, they panic. If you stay composed, focused on problem-solving rather than blame, you give them confidence to do the same. Adaptable leaders create stability not by removing change but by guiding their people through it with clarity and composure. In moments of uncertainty, adaptability means staying calm, pivoting with purpose, and finding opportunities in the unexpected. It's developing the confidence to see challenges as temporary and growth as ongoing. This mindset extends beyond bouncing back, it stretches to leaning in, learning, and using the experience to improve.

In one project launch, I witnessed multiple teams adapted in real-time when setbacks arose. The project involved teams and individuals from three different companies, distributed across the U.S., with some joining as early as 4:00 AM. During the rollout and initial testing, setbacks were encountered. After two hours of troubleshooting, it was concluded that it was best to push the launch. I was incredibly impressed, and grateful, for the team's ability to adapt in real-time, adjust the plan, and keep the project moving forward. I've seen teams in the past fall apart in moments like these, with finger-pointing, blame, and ego taking center stage. That wasn't the case here. Instead, all parties focused on fixing the issues and a launch later in the week.

The ability to navigate uncertainty begins with maintaining control and continues with being ready to embrace the possibilities that come with the unknown. Adaptability has moved beyond flexibility. It's rooted in growth. It's the quiet confidence to face whatever comes your way and make the most of it. Adaptability is how we **Become More** in practice, not simply better at what we already know, but more capable, more resilient, and more prepared for whatever comes next. It's the trait that transforms principles into lived action under real-world conditions.

How to Become More

Adaptability is clarity in change. Use these prompts to reflect on how you reset and rise in moments of uncertainty.

- How do you reset between meetings to ensure your focus, energy, and contributions align with the moment?
- Are you tailoring your message to suit the needs of your audience, or defaulting to what's easiest for you?
- Think about the last time plans fell apart or a challenge emerged. How did you react? Did you resist the change, or rise to meet it?

One MORE Thing

In the modern workplace, adaptability is the defining trait. It sustains every other strength and equips you to move forward no matter the path ahead. To be adaptable is to be ready to contribute at any moment - in calm or in chaos and to see every change not as a setback, but as a chance to grow stronger.

35 | TEMPERANCE - A GUIDING LIGHT FOR BALANCE AND GROWTH

Harnessing the Power of Temperance to Lead

Temperance provides strength through steadiness. It is the practice of self-restraint that brings clarity, not constraint. True discipline isn't limitation; it's the steady restraint that creates space for wiser choices, sharper vision, and sustainable growth. Temperance is best understood as balance, moderation, and mindful self-control. It's the inner virtue that guides all the other traits — not dulling ambition but channeling it wisely. Some confuse temperance with weakness or restraint that stifles drive, but it's the opposite: it gives leaders the clarity to know when to push forward and when to hold steady.

Unlike courage or kindness, which often reveal themselves in bold, visible ways, temperance operates quietly. It shows up in the pause before reacting, the decision to listen before speaking, and the choice to wait when rushing ahead would cause harm. In this way, temperance becomes a stabilizing force capable of grounding leaders in clarity, focus, and intention.

This makes temperance different from balance. Balance is about how we align our lives across competing demands. Temperance is about how we govern ourselves in the moment. It's the inner discipline that keeps ambition from becoming overreach, and emotion from becoming reaction. Amid the relentless pace of today's always-on economy, temperance is not passivity. It's an active discipline, a strength expressed through calm and patience. It steadies us in uncertainty and prevents overreach in moments of ambition or stress, and creates the conditions where growth can take root.

Adaptability helps us flex to change, but temperance helps us endure it. It's the quiet strength that allows leaders to stay the course without burning out themselves or their teams. In this way, temperance is not about holding back. It's sustaining progress for the long haul.

Temperance Through Division and Crisis

As President of the Warrington Soccer Club, I had to lead through the uncertainty of COVID. March is the beginning of our spring season. Parents were anxious, board members were divided, and every decision felt like it carried enormous weight. Tempers sometimes flared, with strong opinions on health, safety, and finances. I realized my role wasn't to "win" an argument but to temper the discussion. By listening, pausing, and guiding us back to shared goals, I helped the board move forward with clear, measured decisions. We cancelled the season and refunded every member. Looking back, temperance was the hidden glue that allowed us to keep the club alive during one of its hardest seasons.

Follow A Founding Father - Benjamin Franklin

Benjamin Franklin understood the importance of temperance so deeply that he placed it first on his list of thirteen virtues[1]. He defined it with clarity:

Eat not to dullness; drink not to elevation.

Franklin saw temperance as the foundation of self-discipline. For him, it was a virtue that enabled clarity of thought and action. He believed that by practicing temperance, one could build the balance needed to pursue other virtues and, ultimately, live a more meaningful life.

As a founding father, Franklin also exemplified the art of discourse, engaging with diverse perspectives and navigating conflict to find common ground. This principle is as relevant today as it was during the birth of this nation. In times of political transition, temperance can serve as a powerful tool for bridging divides. It calls us to listen with empathy, to resist the pull of emotional reactivity, and to seek understanding even in disagreement. Honor difference and dialogue.

Franklin's legacy reminds us that the strength of a community, and a nation, lies in its ability to embrace differing perspectives without losing sight of shared values. As we reflect on this moment of transition in our country, temperance provides a way forward. It challenges us to engage thoughtfully, to balance conviction with openness, and to approach conversations with a genuine desire to learn and connect. Dr. Tavris, co-author of *Estrogen Matters*[2] and *Mistakes Were Made (but not by me)*[3] shared a similar sentiment in an interview that has stuck with me:

Try to avoid all rigid, rigid ideas.

It has led me to simply be more open-minded. Instead of asking "Why?" I have been asking, "Why not?" with temperance.

That's why temperance serves as the anchor of the Execution pillar. Courage starts us, critical thinking sets direction, creativity opens new paths, and adaptability helps us adjust along the way. But temperance is what steadies all of them. It ensures we can **Become More** not just in bursts of effort, but consistently, sustainably, and without losing ourselves in the process.

How to Become More

Temperance is discipline with intention. Use these prompts to reflect on how you bring steadiness into your daily actions.

- Think of a recent situation where you reacted impulsively. How might temperance have shifted your response?
- Consider a relationship in your life where tension or misunderstanding exists. How could temperance help you approach the situation with greater balance and <u>empathy</u>?
- Reflect on how you're navigating the current moment of transition, how can temperance help you remain grounded and intentional in your daily actions?

One MORE Thing

Temperance is strength expressed through balance. It steadies us through change, grounds us in clarity, and creates the space where growth can take root whether for ourselves, someone else, or the team.

BLANK – FOR NOTES

E | EXECUTION

The best way to predict the future is to create it.

— Peter Drucker

Execution is the pursuit of contribution and productivity. It is a bias toward doing, applied to the best use of our time. The MORE Effect reminds us that success is not measured in motion but in meaningful action in what we create, deliver, and bring to life. Fulfillment and actualization lie not in endless tasks, but in the discipline of doing the work and **Becoming More**.

E | EXECUTION

Traits: Courage, Critical Thinking, Decision-Making, Creativity, Adaptability, Temperance.

BLANK – FOR NOTES

36 | MODERN LEADERSHIP BUILT FOR MORE

Protecting What's Human in the Age of AI

Mindfulness. Opportunity. Relationships. Execution.

These four pillars — encompassing 26 human-centric traits — have formed the foundation of your learning so far. Completing all 26 traits is a major milestone in building a leadership approach for the modern world. These traits now stand as your definitive playbook, practices you can return to in any season, any role, any challenge. With them, you are better prepared to lead in a world where the rules have changed, even if the systems around us have not yet caught up.

Take a moment to acknowledge the work you've done. You've examined your character, reflected on your habits, and engaged with core disciplines that set modern leaders apart. That's no small accomplishment. You now hold a complete toolkit — built not for the past, but for today's AI-first, always-on world. This is the new standard of human-centric leadership.

Above all, this playbook asks leaders to protect what's human—our attention, trust, and character—so the tools of AI serve people, not the other way around. Together, these traits form your compass for what comes next: the human disciplines that make effective leadership possible in a rapidly changing workplace.

As Theodore Roosevelt once reminded us,

> *"Let us therefore boldly face the life of strife, resolute to do our duty well and humanly; resolute to uphold righteousness by deed and by word; resolute to be honest and brave, to serve high ideals, yet to use practical methods. Above all, let us shrink from no strife."*

These words tell us to face struggle directly. In that spirit, we now prepare to turn strife into strength, carrying forward what we've built and forging it into the strength of **More**.

37 | THE STRENGTH OF MORE: INTRODUCING THE HUMAN CODE

From Traits to a Code of Living

Now, with your compass in hand, it's time to navigate forward. The next step is to turn these traits into a living system that will guide you through the modernization of how we work, live, and lead: a Human Code to ground you and help you thrive in an AI-driven world. With so much shifting around us, we need more than tools. We need something deeply human to ground us, keep us aware, and keep us moving. Early in the shaping of The MORE Effect, one idea kept echoing:

The endurance of human existence depends on one word: **More**.

It's bold, but it centers a conversation that must be had. In a world racing toward artificial intelligence, machine learning, and automation, the only way we keep up is by becoming more human, not less.

More present.

More thoughtful.

More connected.

More effective.

It is a deeper, broader lens for the application of the 26 MORE Leadership Framework traits. It's a **code of living.**

A Human Code

A principle driven model for thriving in the AI-first, always-on world created to center us on what matters, helping others **Become More**.

A code for how we show up every day.
A code for how we support others.
A code for challenging rigid systems and seeing the world differently.
A code for honoring difference and dialogue.
A code for staying grounded and human.

It's a code for working. A code for leading. A code of living. A way to lead in the modern world, where four generations are now sharing the workplace. Where Gen Z is entering. Gen Alpha won't be far behind. Baby Boomers are transitioning out. And in between, people are navigating hybrid work, return-to-office mandates, caregiving, parenting, eldercare, mental and physical well-being, and the persistent pressure to be always-on. This framework doesn't long for the past. It doesn't insist on rigid models that no longer fit. It meets people where they are and helps them move forward.

It's a blueprint for honoring the *whole person*, not just the employee. For recognizing the role of the *whole family*, not just the professional identity.

Because being a working mother, a caring father, or a present partner isn't a distraction from performance. It's part of what shapes how we show up every day, and why. To ignore that truth is to cling to the past, which is foolish in a world where machines grow more intelligent by the day. This is how we modernize how we work, live, and lead. With more clarity. More care. And more courage.

This isn't a manifesto. It's a **Code of Living**. A **Human Code** for the Age of AI.

A practical, emotionally resonant framework for modern life. A way to name and practice what already exists inside us with more clarity and consistency. This framing is designed to guide not just performance, but how we live and lead. This reframing isn't meant to simply explain what high performance looks like, it's also meant to show what it *feels* like. It gives us a way to lead with intention, not reaction.

The Four Codes That Power Human Potential

The **MORE Leadership Framework**'s four pillars and 26 human-centric traits form the foundation of modern leadership—a complete system for how humans thrive today. When we view them through this lens, leading and thriving in the AI-first, always-on world, where individuals are seen and heard, each pillar reveals its emotional purpose. The why beneath the what. Each offering its own influence into a code of living, a human code.

M | Mindfulness → The Code of Presence

This is the foundation. The ability to be in the moment, not scattered, present. To respond, not react. To be calm. It's where self-awareness becomes skill. Where stillness meets precision. In a world that demands our attention, **Presence is our power**.

O | Opportunity → The Code of Possibility

This is vision. The will to carry on, accepting that to live is to struggle. The belief that better is real and available. That setbacks set up the next opportunity. That failure is fine. That there's always something to build, to learn, to try, turning fear into curiosity, and making progress feel personal. **Possibility is our strength**.

R | Relationships → The Code of Connection

This is where leadership becomes human. Where accountability meets care. Where we look each other in the eye and say, "I've got you." Needing others isn't soft. It's human. Trust only scales through kindness and generosity. **Connection is our currency**.

E | Execution → The Code of Contribution

This is productivity with meaning. Not just doing more, but doing what matters, and **Becoming More**. It's consistency over chaos. Energy over ego. Intent, action, and character lead to outcomes and shape the story we live. **Contribution is our legacy**.

Why This Matters Now

We're working, living, and leading in a rapidly evolving culture shaped by new technology, new motivations, and new rules being rewritten. The pace has shifted dramatically. The pressure has increased. And people are asking deeper questions. Not only, "How do I succeed?" but also "How do I find my purpose?"

That's why this framing matters. It gives us a language to name the energy behind performance and helps leaders build cultures that are **both fast and fair**. It gives teams something real to rally around. It helps parents model resilience. It helps friends show up better. And most of all it gives us a path to grow without losing ourselves. Because ultimately, this is more than a leadership tool, it's a reclamation of self. A framework for agency. A modern blueprint to improve the human condition in a world that's being reshaped in real time. You don't have to trade pace for presence. You don't have to give up connection to get results. We need a system built for both.

This **Human Code** – The MORE Effect – is that system. Not just a framework for leadership, but a blueprint for human endurance.

What's Next

Now, we arrive at a turning point. This moment marks the shift. From traits to system. From a list to a living lens on life. In the chapters ahead, we'll build on this foundation bringing this reframing to leadership and daily life. This throughline is the bridge between emotion and execution, between potential and proof. If we want to lead today:

We must understand what makes us human — and never lose sight of it.

BLANK – FOR NOTES

SECTION 3

THE HUMAN CODE
FOUR CODES OF MORE

Earlier I shared how my mother inscribed Courage. Hope. Optimism. Pursuit of all three in a book to encourage me through a difficult time. These words have been my compass. They remain the through-line of this framework. Now, they carry us into what's next. You understand the traits of the **MORE Leadership Framework**; in this section, they come to life inside four codes.

- **From Mindfulness to Presence**
- **From Opportunity to Possibility**
- **From Relationships to Connection**
- **From Execution to Contribution**

This is where the framework becomes practice — no longer abstract, no longer aspirational. It is the Human Code for the modern world: a system to thrive in the future without losing what makes us human. Use each code as a weekly focal practice: pick one, apply the prompts, and notice what changes.

BLANK – FOR NOTES

38 | THE CODE OF PRESENCE - FROM DISTRACTION TO ATTENTION

The Discipline of Focus

The average American spends more than five hours a day on their phone[1]. What's left for presence? Presence is a lifeline: it converts time into connection and keeps trust alive. We don't have a shortage of attention. We have a shortage of *presence*. In an AI-first, always-on world, presence isn't a luxury. It's a lifeline. It's the human code we come back to when everything else pulls us away. We check our phones without thinking. We scroll feeds without noticing. We respond to notifications mid-conversation. And then we wonder why we feel fragmented, disconnected, behind.

I won't tell you to quit technology. I use my phone constantly: for work, for writing, for building, for creating. But I've also caught myself missing moments that mattered. I've fallen into the scroll when I meant to stay grounded. Just like everyone else. Again, on average we now spend over five hours a day on a phone. That may not sound like much until you realize it adds up to **almost a full day and half every week**. Thirty-five hours, gone. That's time we could've spent looking someone in the eye instead of watching their stories. Time we could've used to listen, to think, to connect. Presence isn't anti-tech; it's pro-human. It is self-awareness applied—an honest look at how you're spending your time. And self-awareness is the starting point for real growth.

A newly married friend asked me for relationship advice. I've been married since 2001. My advice was simple: **Show up. Be present.** That's the real work of a relationship. Not always the big gestures but the everyday quiet, consistent care and attention. It's noticing when your partner needs you and being there without being asked. And you can't do that if your head is buried in your phone.

Presence turns time into connection. It keeps trust alive. It makes love visible. Without it, none of the other codes work. You can't contribute in chaos, create connection through distraction, or see possibility through noise.
Presence isn't passive. It's active. It's a choice. And it's available at the same moment you reach for your phone. Consider: What if you gave your presence the same attention you give your phone? What if, as easily as you open Instagram or email, you opened a moment of stillness? What if you gave someone your full attention not because they demanded it, but because they *deserved* it? And what if you offered yourself that same gift?

Not perfection. Not performance. Just presence.

That's where **More** begins.

To close, I want to be clear. Sometimes writing feels like journaling. I struggle with presence too. I often find myself deep in thought, on my phone or not, and missing the moments right in front of me. With writing, I'm able to think through my thoughts, helping my own self-development and ability to show up. I don't write them because I've figured it all out. I write them because I'm still figuring it out, just like you. The fact is doom-scrolling and phone addiction are real. Don't let your device control you.

Being more present in all phases of my life is something I'm working on. The more I pay attention to it, the more I see how much it matters.

And I hope this advice offers you something to think about as you step into your days, with a little more clarity, and a little more presence.

How to Become More

- If your presence feels fractured — at home or work — what one thing could you do to make it feel less so?
- Where does your attention go when you're overwhelmed or distracted?
- What would it look like to reclaim one hour this week and give it fully to someone else, or to yourself?

One MORE Thing

Attention, freely given, is the greatest gift within your control. Give freely.

39 | THE CODE OF POSSIBILITY - FROM DOUBT TO HOPE

The Discipline of Belief

I didn't fully understand possibility until I became a father.

In that moment, tiny hands, blinking eyes, no words, just awe, I became both protector and nurturer. Suddenly, someone else's **possibility** was my responsibility. A new life was bestowed upon us and in that moment as a parent you realize the weight of holding their possibility for them until they could carry it on their own. That's what leaders do too. They carry vision. They keep going even when life or work feels like struggle. They believe that better is possible and remind others that setbacks can act as setups and that failure is fine. Possibility doesn't ignore fear — it turns fear into curiosity and motivation, making progress feel personal.

You don't just hold the vision for the business. You hold belief for your team, your mission, and your people. You help others see in themselves what they can't yet see. And in doing so, you give them the greatest gift a leader can offer: the transfer of belief. Possibility becomes real the moment someone is willing to believe on your behalf.

Possibility isn't passive.

It isn't idle optimism or blind hope. It's a decision. A daily act. A muscle. Possibility says: *we can be more effective, more productive, and we can be more together,* even when we don't know exactly how. It insists on trying again when the last plan didn't work. It steadies the hand that reaches forward when it is easier to fold. And it matters most when energy is low, when progress is slow, when fear begins to take up too much space. Because once belief dies, behavior follows. And when skepticism hardens into cynicism, momentum breaks and culture suffers. Cynicism drains the room. It shuts down the second question. It strips the connection of warmth. It turns collaboration into survival mode. You see this in companies when hallway conversations disappear. When silence becomes safety. That's not just a performance problem.

That's a belief problem.

And belief isn't restored by pretending. It's restored by re-grounding.

The slump is not the enemy.

In any high-performing system, whether a team, a startup, or a season, there will be dips. It's inevitable. And yet, far too often, we treat these dips and slumps as systems failure instead of signals to recalibrate. But a slump is like a batting cold streak. You don't rewrite your swing. You return to the fundamentals. The stance. The breath. The rhythm. The belief.

You remember what you're building. You anchor back to your values. And you remind each other: we are not defined by this moment, but by what we do next. Because that's what possibility offers: not escape, but clarity.

Possibility's power is in reframing.

It doesn't lie.
It *sees.*

It sees what's broken and calls it a test.
It sees failure and finds feedback.

It sees stagnation and asks for space, not shame.

Possibility turns "We're losing" into "We're learning."

It turns "This is broken" into "This is testing us."

It turns "We don't know what to do" into "Let's go back to what we *do* know and start there."

It turns "That person isn't ready" into "They haven't been *seen* yet."

It turns "I'm stuck" into "I'm in between chapters."

It turns "This isn't working" into "This needs attention."

Reframing is not spin. It's leadership. It's the lens that shapes your landscape. **The story you tell becomes your reality**.

Possibility is fueled by generosity.

Possibility doesn't guarantee progress. Sometimes it doesn't move us forward, it simply steadies us while we figure out what's next. But it does keep open the path when energy fades or belief wanes. It turns doubt into the next step. It reminds us that progress is still possible, even in the slump.

That's why generosity matters. Belief given away multiplies. When you lend hope to your team, your family, or even yourself, you control the story. A slump becomes a reset. A setback becomes a signal. Offering a pause to ask yourself, "How are we so sure this is such a bad thing?"

Possibility restores the ground beneath your feet. This is how to build forward. This is how to honor the moment. This is how to be responsible for hope. This is how to unlock the full potential of optimism and opportunity, by turning belief into action, and action into contribution.

Possibility is awe. I learned that holding my daughter for the first time. Her tiny hands, blinking eyes, and fragile breath reminded me that every beginning is a miracle — and that belief is the bridge from fear to future.

Every parent feels it in that first moment: the shock that life can be this fragile and this full of promise at the same time. That same awe is what leaders are called to hold for their people. To carry their possibility until they can carry it for themselves.

As Goethe reminds us:

Whatever you can do, or dream you can, begin it. Boldness has genius, power, and magic in it.

That's the Code of Possibility.

How to Become More

- Where in your work or life has cynicism begun to replace belief? How can you reframe it?
- Who around you needs you to believe in them until they can believe in themselves?
- What slump are you facing right now, and what "fundamentals" can you return to?

One MORE Thing

To live is to struggle. To lead is to believe. Let possibility be the awe that steadies you, the sense of wonder that reminds you progress is still possible, even when the path isn't clear.

40 | THE CONNECTION CODE - FROM CURIOSITY TO BONDS

The Discipline of Kindness

The road to **More** starts with one warm look, two curious words, and a small act of generosity.

Helping every moment Become More.

At a large conference, I decided to run a small experiment: could one gentle smile start a chain reaction of help inside a cavernous expo hall? My only rule for the week was simple: every time I made eye contact, I smiled first, asked "Tell me...," and then offered one small act of service before we parted. Three days later, after significant time outside my comfort zone (I feel safest as an introvert), I met at least a hundred people and a few with whom in a few minutes of conversation a friendship developed. What I learned was not new, but it was newly confirmed.

Connection compounds. It begins with three human moves anyone can master. Rick and Terry modeled the idea while mentoring me; Zig Ziglar introduced the principle to the universe:

You can get everything you want in life if you just help enough other people get what they want.

That single sentence is the sentiment that threads through everything that follows. It is also the heart of The MORE Effect. We endure and thrive by choosing to **Become More** for one another. Selflessly.

I call the practice **The Connection Code**.

Three small-behaviors, executed in order, transform ordinary exchanges into shared moments, helping interactions **Become More**.

The Path To Connection

I've found that today, when we stare at our phones nearly five hours a day, humans have become increasingly self-absorbed. The pandemic has altered our nature, making us more isolated and guarded. Loneliness and isolation have become the silent quiet causes of sadness and depression.

But as social creatures, humans need connection. Even just a fleeting two- to three-minute unexpected conversation can shift one's day. After some experimentation in expo halls, restaurants, trains, I landed on a simple formula for making connections.

1 | Kindness -> Smile
2 | Curiosity -> Ask
3 | Generosity -> Give

The Connection Code

1 | Kindness

The fastest path to kindness is soft eye contact with a gentle smile. A warm look says, *"I see you; you're safe here."* Neuroscience tells us that eye contact lowers cortisol and opens the social learning centers of the brain. In practice it does something simpler: it offers comfort. Whether you are greeting a friend or a stranger: meet eyes first, smile second, speak third. Reading this moment for openness is essential, as not every moment is ready for **More**. You'll know when it is.

2 | Curiosity

With the unspoken agreement found in eye contact and a shared smile, become curious. The easiest way starts with two simple words: **"Tell me..."**

- Tell me, where are you heading? Tell me, where are you from?
- Tell me, who do you work for? What do you do?
- Tell me, where did you go to college?
- Tell me, what has you most interested at this event?

Curiosity suspends judgement and invites story.

3 | Generosity

Your intentional presence and attention are the first gifts you give the moment through active listening. Listen with intent to serve, to help the moment **Become More** for the other person. Within minutes a shared thread or interest may emerge; an experience, a location, a person, an event, a podcast. Follow that thread and it will reveal what the moment needs you to give. You're not seeking anything profound, a simple act of generosity is enough. Remember, *If you just help enough other people get what they want ...* And if nothing obvious emerges, it's possible the only thing the moment needed was your full attention.

Connection rarely ends with a handshake; it leaves a remnant of possibility. Sometimes it sparks immediate collaboration; other times it just lightens a stranger's load. Either way, The Connection Code ensures you leave people feeling better than you found them — and that is how The MORE Effect scales encounter by encounter, moment by moment, person by person. Every act of seeing, asking, and offering what the moment needs widens connection and proves that **More** is available: right here, right now, in the Age of AI.

Putting the Code to Work

Connection isn't a talent; it's a choice.

Kindness opens the door. Curiosity invites the story. Generosity carries it forward.

From Moments to More

Practiced in sequence, these small acts turn ordinary moments into something **More. That's The MORE Effect**. Connection doesn't require charisma or grand gestures, just intent. A warm glance, two curious words, and one act of help reset the energy of any interaction. Use the **Code of Connection** in hallways, video calls, and coffee lines. You'll notice moments <u>feeling</u> instantly different and better ... **Becoming More**.

In the Age of AI, these acts of kindness both simplify and amplify human connection. Each small deposit proves that **More** is available, transforming chance encounters, and even your closest relationships, into possibility.

Reflective Prompts

- When did someone's warm glance last melt your guard?
- Whose ideas have you only half-heard? What could those simple two words, "Tell me..." reveal?
- Where could a quick intro or note lighten someone else's load?

One MORE Thing

Your responsibility: leave people <u>feeling</u> better than you found them.

41 | THE CODE OF CONTRIBUTION - FROM MOMENTS TO LEGACY

The Discipline of Effectiveness

Did I do my best today?
Did I give the moment more?
Did I contribute to the greater good?
Did I protect my well-being so I can keep contributing tomorrow?

These are the questions at the heart of Contribution.

Because how you spend your time is the clearest measure of effectiveness. Effort without direction wastes time and money. Busyness without purpose is unfulfilling. Skating by is low character. And work ethic is a reflection of integrity and discipline.

Contribution is the code that transforms presence, possibility, and connection into something lasting. It is execution with purpose, effectiveness with endurance. It's the discipline of making sure the work you do, the way you show up, and the energy you spend all add up to more than activity — they add up to legacy.

The Discipline of Contribution

Contribution demands discipline: the discipline of direction, the art of action, the prioritization of time, and the restraint to know when to pause — and when not to.

We often picture contribution as big, dramatic breakthroughs. In reality, it is usually the steady, faithful effort of high-character people doing a good day's work—small steps toward meaningful outcomes. For knowledge workers, effectiveness is the most important thing you can deliver, regardless of level or title.

True contribution is where function and fulfillment meet. Contribution isn't limited to the office or the scoreboard. It lives in life and work, in big projects and small interactions. Every encounter leaves an impression — positive or negative. You'll always make an impact, one way or another. The real question is simple: How do you want to be remembered? It's not only delivering outcomes — but also knowing that what you delivered mattered. How will you be remembered? The effectiveness of your effort will be the story of your contribution.

Time and Scarcity

There will always be more contributions to make than hours in a day. Time is scarce. That's the human condition. Prioritization isn't optional. **The critical question is: What in your work deserves priority?** Contribution asks us to choose carefully, because not everything deserves our energy. Steady often beats fast. Effectiveness and productivity aren't about chasing everything; they're about pursuing the few things that matter most: measured against the deadlines and objectives in front of you.

Effectiveness Over Busyness

Contribution is not about doing more; it's about doing what matters. Being productive doesn't mean being busy — it means achieving meaningful results. It's not how hard you're working; it's how smart. Peter Drucker said it best:

> *"The focus on contribution is the key to effectiveness."*

Effectiveness is not a trick, but a habit. It must be learned. Contribution becomes instinctive when you train yourself to ask every day: What is the highest and best use of my time? Where can I make the greatest difference? How can I be certain my team understands how to best apply their time? This is the work of the modern leader.

The Modern Work Ethic — Doing Work That Matters

In an AI-first world, we have an abundance of intelligence at our fingertips. Forty hours of knowledge work today should produce more than it did a decade ago. That makes it even more urgent to do work that matters — outcomes aligned with purpose, not just output. Don't waste energy on busywork that can be automated.

The critical skill of contribution is no longer output; it's discernment.

Discernment is choosing the highest-leverage work, sequencing it well, and saying no to the rest. It's how you protect your time and direct your effort toward what moves the mission forward and often the simplest way to know if you chose well is how you feel at the end of the day.

Making yourself proud matters. Your achievements reflect your character. What do others see? What do they learn from your actions — not your words? What do you do when no one else is looking? Are you proud of it? Each day, give yourself the mirror test: if you're not proud of what you contributed, if it wasn't your best work, ask yourself why. Don't just rethink how you're spending your time, seek the real answer. Your job won't always align perfectly with purpose or passion. But that doesn't excuse you from contributing an honest day's work.

A good day's work is more than checking boxes. Contribution is choosing to spend your finite time on what creates value — for yourself, your team, and the world — no matter the situation you find yourself in.

From Moments to Legacy

Contribution extends beyond tasks into relationships, health, and community. AI may take on busywork, but only humans can offer presence, belief, and connection. Legacy is built not by what you got done, but by what you gave away. What you give in moments adds up to your legacy. Legacy is not only what you built, but the lives you made better along the way.

Legacy is built moment by moment, encounter by encounter.

The MORE Effect in action as a code of living, a human code.

In the end, contribution is your integrity.

Your character.

Your legacy.

The more effectively you use your time, the more you contribute and produce — at work and for humanity — the prouder you will be of the legacy you leave.

This is the Code of Contribution.

How to Become More

- Where are you confusing activity with contribution?
- What of your work deserves priority this week?
- When was the last time you felt truly proud of your contribution?

One MORE Thing

Contribution is not simply doing more. It's doing what matters most — moment by moment, encounter by encounter, until it becomes your legacy.

BLANK – FOR NOTES

THE HUMAN CODE – A CODE FOR LIVING

42 | CARRYING ON

The Discipline of More, Every Day

You've now completed the four Codes.

Presence, Possibility, Connection, and Contribution.

They are designed to elevate our attention to character and reclaim the values and ideals - not lost - but fading in our society. Their purpose in The MORE Effect is to instruct us for the Age of AI — providing **A Human Code** for the modern world — a map for the artificially intelligent future still being written.

But even the best systems need orientation. In a landscape where technology accelerates every decision and AI reshapes how we work and live, it's easy to drift. That's why, before we close, I want to leave you with something simpler: five guiding principles.

Think of the Codes as disciplines you *practice*. Think of the Principles as anchors to keep you on course. They are reminders you carry to stay grounded, true, and human.

They are not rigid rules. Simple reminders. Again anchors. To keep you from drifting. A way of asking the right questions in the moments that matter most.

These are the **Principles of MORE**.

BLANK – FOR NOTES

43 | THE PRINCIPLES OF MORE

Five Anchors for Thriving

In an age defined by artificial intelligence and constant digital demands, staying true to what makes us uniquely human has never been harder. As we navigate this new world, the **Principles of MORE** serve as our guardrails. They are five simple, memorable rules.

To keep us present.

To keep us human.

To help us **Become More**.

The principles are a commitment: that **Time Is All You Have**, a reminder to **Protect What's Human**, a call to **Give the Moment More**, a pledge to **Honor Difference and Dialogue**, and a discipline to **Not Let the Device Control You.**

These rules are here to help you lead and live with intention.

They're simple, memorable, and for the future still being written. Anchor yourself here. You'll find it easier to stay true to your values and guide others to do the same.

Principle 1: Time Is All You Have

You get one life. How you spend your time is all you have. In a world that constantly pulls at our attention, this principle reminds us to pause and ask: *Is this a good use of my time?* Make every moment matter by investing your hours in what truly enriches you and those around you.

Principle 2: Protect What's Human

Artificial intelligence can do many things, but it will never replicate the depth of human emotion. This principle is about safeguarding the qualities that make us uniquely human. Leadership isn't just about efficiency; it's about humanity in every decision.

Principle 3: Give the Moment More

We endure and thrive by choosing to give generously to each interaction — more attention, more kindness, more listening. Make each moment count by contributing positively to the world around you. When we add more — attention, kindness, care — we help one another, strengthen our community, and create a lasting sense of belonging.

Principle 4: Honor Difference and Dialogue

In a diverse world, honoring difference is leadership at its highest character. This principle calls us to respect varied perspectives, engage in dialogue, and embrace civility. Recognize that our differences are a source of strength and unity, not weakness and division.

Principle 5: Don't Let the Device Control You

Phone use is fine. Phone addiction is not. Technology is a tool, not your master. This principle is a reminder that in the age of AI, the algorithms seek control. Use your phone to connect, create, and learn — but notice when it starts shaping your thoughts or consuming your time in ways that don't serve you and your values.

Bringing MORE To Life

By anchoring yourself to these principles, you will find it easier to stay true to your values and guide others to do the same. Because in the end, **becoming more** is not a single event — it is a daily choice.

It begins, and begins again, with these small steps.

Choose More.

Become More.

The Principles of MORE

1 | Time Is All You Have
2 | Protect What's Human
3 | Give the Moment More
4 | Honor Difference and Dialogue
5 | Don't Let the Device Control You

BLANK – FOR NOTES

44 | BECOMING MORE IN THE AGE OF AI

Courage. Hope. Optimism. Pursuit of all three.

We are alive in a remarkable time.

A time where artificial intelligence can mimic conversation, compose music, write code, and solve problems faster than the human brain can process them. A time where we stand on the edge of Artificial General Intelligence (AGI), when machines may rival human capacity across domains. A time where tools that once took decades to develop now arrive in days. Where we no longer wait for the future. It shows up in our inbox.

And yet...

We're also living in a time of loneliness. Burnout. Division. Distrust. Rapid acceleration without clear direction. High output, low meaning. Constant connection, yet we feel unseen.

This is the tension of the always-on world: more connected than ever, and yet more disconnected from ourselves.

In the previous chapters, you uncovered the answer. The answer, which you've built over the previous chapters, is The Human Code: a system for thriving in the future without losing what makes us human.

From Distraction to Attention - The Discipline of Focus
From Doubt to Hope - The Discipline of Belief
From Curiosity to Bonds - The Discipline of Kindness
From Moments to Legacy - The Discipline of Effectiveness

We need a new way to live. A new way to lead. A new way to become.

That's what The Human Code is. A Code for Living in the age of AI. It is not a rulebook. It's not a checklist. It's a framework for staying fully human in a world that's being rapidly reshaped by machines. It's a code for how we show up at work, at home, in the quiet moments in between. It's how we respond to complexity with clarity. To speed with stillness. To noise with intention. To fear with courage. To uncertainty with care.

It is a reclamation of self.

A rediscovery of leadership.

A redefinition of what it means to **Become More**.

This is **The MORE Effect** fully alive.

But a code is only as good as its application in moments of pressure. So when you find yourself drifting, when the noise of the world becomes too loud, this is how you use your code.

First, ask the right questions:

- Am I present for what matters?
- Am I believing in what's possible or defaulting to comfortable?
- Am I building connection, not just communication?
- Am I contributing with intention or just moving?

If the answers are unclear, return to your four Codes:
Presence, Possibility, Connection, and Contribution. They are your compass.

And if you need grounding, anchor yourself in the five Principles:
Time Is All You Have, Protect What's Human, Give the Moment More, Honor Difference and Dialogue, and Don't Let the Device Control You.

This is your direction.

This is your edge.

This is the Human Code of The MORE Effect.

A code of living for thriving in an AI-first, always-on world. A framework for becoming when everything around you is accelerating. Artificial intelligence might soon outthink us. It might out-optimize, out-analyze, even out-predict.

But it cannot do this:

- It cannot wonder.
- It cannot wander.
- It cannot forgive.
- It cannot feel awe.
- It cannot show restraint when ego says act.
- It cannot create belonging.
- It cannot hold space.
- It cannot lead with heart.

That final point—**the ability to lead with heart**—is the essence of this code. The ability to care deeply, reflect honestly, and grow intentionally.

These aren't soft skills.

They are survival skills.

And they are the reason we endure.

I learned this lesson before I had the words for it. My mother saw it in me before I saw it myself. She inscribed a book with the four words that would come to define my life:

Courage. Hope. Optimism. Pursuit of all three.

Four words. One nickname. Chop.

That name became my reminder: that even in the hardest moments, we're born to overcome, we're meant to keep becoming. To believe. To carry on. To help others do the same.

You get one life. How you spend your time is all you have.

So, spend it **Becoming More**.

Spend it giving **More**.

Spend it living **More**.

For you.

For your family.

For your community.

For humanity. Because,

The endurance of human existence depends on one word: **More**.

EPILOGUE

KEEP BECOMING MORE

The end of this book is not the end of the work. It is the beginning of a practice. Courage. Hope. Optimism. Pursuit of all three.

Those words, inscribed for me long ago, remain the heart, soul, and inspiration of The MORE Effect. They remind us that becoming more is not a single decision or a single season. It is a lifelong struggle. The Human Code isn't a checklist or a set of rules. It's a system to return to when the world feels overwhelming, when AI races ahead, when your own energy runs thin.

Presence. Possibility. Connection. Contribution.

These four codes are anchors, reminding us that while machines may think faster, only humans can live with intention. You do not have to master all twenty-six traits tomorrow. You do not have to be unshakable every day. You only have to keep *becoming*. One choice. One conversation. One act of kindness, courage, or curiosity at a time. Life and leadership are not about being perfect. They're about being willing. Willing to examine yourself. Willing to help others grow. Willing to hold on to what is uniquely human, even in an AI-first, always-on world.

If there's one truth to carry forward, it is this: you were not meant to keep pace with machines. You were meant to be a part of your family. You were meant to lead people. To steady them with presence, to inspire them with possibility, to connect them with care, and to move them forward with contribution. That is how we endure. That is how we thrive. That is how we create a future worth belonging to. So, wherever you find yourself tomorrow — in a boardroom, on a field, at your kitchen table, or simply alone with your thoughts, remember: the work is never finished. The call is always the same. To overcome. To endure. To reach higher. To thrive.

To **Become More**.

BLANK – FOR NOTES

The unexamined life is not worth living.

-Socrates

You get one life.

How you spend your time
is all you have.

-Jeremy Victor

BACK MATTER

ACKNOWLEDGMENTS

To everyone who has helped me **Become More**,
in life, in work, and in the writing of this book,

Thank you.

I'm forever in your debt.
In return, you have my
promise to help others
Become More.

ABOUT THE AUTHOR

Jeremy Victor is a futurist, builder, and senior executive whose career spans over 30 years across technology, digital health, and customer experience. Most recently, he served as Chief Customer Officer at Noom, leading one of the industry's largest AI-driven transformations — scaling operations while preserving what makes business human.

Jeremy is the inventor of MoreScore.ai, a patent-pending system that quantifies emotional equity and brand affinity, and the creator of Experience Manufacturing™, a framework for building human, emotionally resonant experiences into any organization. Through Make Good Holdings — home to MoreScore X P E R I E N C E, an AI consultancy for the Age of AI (www.morescoreX.com), theNXT250, and TavernTALX — he consults, speaks, and builds programs that help people and organizations become more — more resilient, more profitable, more connected, and more human.

Grounded in the small town roots of Lansdale, Jeremy lives in Bucks County, Pennsylvania with his wife, three children, and dogs, Rudy and Lexi. Outside of work, he is a youth soccer coach and volunteer, and finds strength in rucking, the outdoors, and Pearl Jam. His friends call him Chop, you can too.

Through his writing, speaking, consulting and the *Business at the Speed of AI* newsletter and podcast, (https://jeremyvictor.substack.com/), Jeremy equips leaders with a new playbook for the modern age. **The MORE Effect** is both his philosophy and his invitation: a human code for thriving in a world shaped by technology but defined by people.

www.jeremyvictor.com

BRING THE MORE EFFECT TO YOUR TEAM

Bulk Sales for Teams and Organizations

Equip your entire team or company with copies of **The MORE Effect**. Bulk purchase discounts are available to support off-sites, onboarding programs, and leadership academies.

Keynotes & Corporate Speaking Engagements

With the **MORE Leadership Framework,** every presentation is customized to meet the audience where are on how to address the most urgent and pressing issue of the day. From intimate executive retreats to large-scale keynotes, these sessions help leaders modernize systems, scale trust, learn how best to advance the outcomes and build collectively intelligent workforces and organizations.

AI EDUCATION – BecomeMORE Academy

Practical, story-driven exercises bring the 26 leadership traits to life, helping teams strengthen presence, opportunity, relationships, and execution together. These workshops create lasting bonds and new momentum across leadership teams, boards, and organizations.

Gatherings, Immersive Experiences, and Intelligence Summit Series Events

Join Jeremy at the *Business at the Speed of AI Intelligence Summits* Immersive gatherings where executives, innovators, and thought leaders explore the future of leadership, customer experience, and organizational transformation.

To work with Jeremy, explore bulk orders, team workshops, or speaking engagements, email Jeremy at: **jv@jeremyvictor.com**

NOTES & REFERENCES

Chapter 9: The New Frontier of Work-Life Balance

1.   McKinsey & Company. (2022, June 23). *Americans are embracing flexible work—and they want more of it.* McKinsey & Company. https://www.mckinsey.com/industries/real-estate/our-insights/americans-are-embracing-flexible-work-and-they-want-more-of-it

Chapter 10: Mindfulness – Building Emotional Intelligence by Staying Present

1.   Jeremy Victor, Instagram, https://www.instagram.com/jeremyvictor

Chapter 12: Emotional Intelligence in Modern Leadership

1.   James MacGregor Burns, Leadership. New York: Harper & Row, 1978.

2.   Walter, F., Cole, M. S., & Humphrey, R. H. "Emotional Intelligence: Sine Qua Non of Leadership or Folderol?" *Academy of Management Perspectives*, 25(1), 45–59 (2011). https://doi.org/10.5465/amp.25.1.45

Chapter 13: Gratitude in Action: Honor Those Who Serve

1.   U.S. Department of Veterans Affairs. PTSD: National Center for PTSD. Accessed 2025. https://www.ptsd.va.gov/

2.   U.S. Department of Veterans Affairs. *Moral Injury.* PTSD: National Center for PTSD. Accessed 2025. https://www.ptsd.va.gov/professional/treat/cooccurring/moral_injury.asp

3.   Morganstein, J. C., Benedek, D. M., Ursano, R. J., & Russell, R. K. "Operator Syndrome: A Unique Constellation of Medical and Behavioral Health-care Needs of Military Special Operation Forces." *Military Medicine*, 185(9–10), e1501–e1509 (2020). https://pubmed.ncbi.nlm.nih.gov/32052666/

Chapter 17: Optimism - The Fuel of Forward

1.   Graham Rapier. "13 Quotes From Bosses Who Mocked Technology and Got It (Very) Wrong: Technology always finds a way to prove its critics wrong." Inc.com, June 15, 2017. https://www.inc.com/business-insider/boss-doesnt-understand-technology-mocks-trend-wrong.html

Chapter 28: Leadership Doesn't Start with Strategy. It Starts with Sincerity.

1.   Niederhoffer, Kate, Gabriella Rosen Kellerman, Angela Lee, Alex Liebscher, Kristina Rapuano, and Jeffrey T. Hancock. 2025. "AI-Generated 'Workslop' Is Destroying Productivity." *Harvard Business Review*, September 22, 2025. https://hbr.org/2025/09/ai-generated-workslop-is-destroying-productivity

Chapter 35: Temperance - A Guiding Light for Balance and Growth

1.   Franklin, Benjamin. *The Autobiography of Benjamin Franklin.* Edited by Leonard W. Labaree et al. New Haven: Yale University Press, 1964.

2.   Tavris, Carol, and Avrum Bluming. *Estrogen Matters: Why Taking Hormones in Menopause Can Improve Women's Well-Being and Lengthen Their Lives — Without Raising the Risk of Breast Cancer.* New York: Little, Brown Spark, 2018.

3.   Tavris, Carol, and Elliot Aronson. *Mistakes Were Made (But Not by Me): Why We Justify Foolish Beliefs, Bad Decisions, and Hurtful Acts.* New York: Mariner Books, 2007.

Chapter 38: The Code of Presence - From Distraction to Attention

1.   HarmonyHit. *Phone Screen Time Statistics: 2025 Data on Average Use.* Accessed 2025. https://www.harmonyhit.com/phone-screen-time-statistics/

BLANK – FOR NOTES

An idea, like a ghost ... must be spoken to a little before it will explain itself.

-Charles Dickens

The explanation is that he has a twenty-track mind, and all the tracks converge on one objective, which is to do something worth while.

-William J. Cameron describing Henry Ford

To Helping Others Get What They Want

Jeremy's spent a lifetime thinking about ideas. Founded in 2009, with an eye toward the future, **Make Good Holdings, LLC** is the playground for Jeremy's ventures. He's about to come out of the sandbox and start building from the ground up - for **#theNXT250** years of American Revolution in the Age of AI.

->   aJV Venture   <-

B2Bbloggers.com ....retired.

Make Good Media Brand, content, media, events, publishing .... Active.
     Books | The MORE Effect ....October 3rd, 2025.
     Podcast | Business @ the Speed of AI ....sign up at: SpeedofAIPodcast.com.
     Prints, Stock Photography..coming 2026.
     Events | Intelligence Summits ....coming 2026.

iShootYouthsports.com -.......................... Sports Photography Franchise ...... coming 2026.

My [STEALTH]: [----------------S T E A L T H ----------------] ....incubating.

MoreScore X P E R I E N C E ....www.morescorex.com.
a consultancy for the Age of AI ... From Legacy to Future Built.

MoreScore.ai ....patent-pending Experience Manufacturing™ platform.........coming 2026.

The NXT 250 - Business Built for America250 - www.theNXT250.com ....coming 2026.
     Living Historical Experiences ....TavernTALX.com.
     Apparel .... shop.NXT250.com.
     Travel & Leisure .... AmericaWALX.com ....coming 2026.

BLANK – FOR NOTES

BLANK – FOR NOTES